INSTRUCTOR'S RESOURCE MANUAL

Robert W. Christopherson

Charles E. Thomsen
American River College

FIFTH EDITION

ELEMENTAL GEOSYSTEMS

Robert W. Christopherson

PEARSON

Prentice
Hall

Upper Saddle River, NJ 07458

Editor-in-Chief, Science: Dan Kaveney
Acquisitions Editor: Jeff Howard
Associate Editor: Amanda Brown
Executive Managing Editor: Kathleen Schiaparelli
Senior Managing Editor: Nicole M. Jackson
Assistant Managing Editor: Karen Bosch Petrov
Production Editor: Debbie Ryan
Supplement Cover Manager: Paul Gourhan
Supplement Cover Designer: Christopher Kossa
Manufacturing Buyer: Ilene Kahn
Manufacturing Manager: Alexis Heydt-Long

© 2007 Pearson Education, Inc.
Pearson Prentice Hall
Pearson Education, Inc.
Upper Saddle River, NJ 07458

The author and publisher of this book have used their best efforts in preparing this book. These efforts include the development, research, and testing of the theories and programs to determine their effectiveness. The author and publisher make no warranty of any kind, expressed or implied, with regard to these programs or the documentation contained in this book. The author and publisher shall not be liable in any event for incidental or consequential damages in connection with, or arising out of, the furnishing, performance, or use of these programs.

Printed in the United States of America

10 9 8 7 6 5 4 3 2 1

ISBN 0-13-154792-5

Pearson Education Ltd., *London*
Pearson Education Australia Pty. Ltd., *Sydney*
Pearson Education Singapore, Pte. Ltd.
Pearson Education North Asia Ltd., *Hong Kong*
Pearson Education Canada, Inc., *Toronto*
Pearson Educación de Mexico, S.A. de C.V.
Pearson Education—Japan, *Tokyo*
Pearson Education Malaysia, Pte. Ltd.

Contents

Outline

Let us begin the dialogue right now. Please consider us an E-mail away if any questions, comments, suggestions, or other inquiries arise. All books and learning resources improve through such feedback, interactions, and critical response.

Our address information:

Charlie Thomsen
E-mail: thomsec@arc.losrios.edu
Home Page: http://ic.arc.losrios.edu/~thomsec
American River College
4700 College Oak Drive
Sacramento, CA 95841

Robert W. Christopherson
E-mail: bobobbe@aol.com
Home Page: http://www.prenhall.com/christopherson
P. O. Box 128
Lincoln, CA 95648-0128

Introduction

For Prentice Hall:
geosciences@prenhall.com
http://www.prenhall.com
USA: 1-800-374-1200

phcinfo_pubcanada@prenhall.com
Canada: 1-800-567-3800

Physical Geography in the 21st Century

Welcome to physical geography! Thus begins the fifth edition of *Elemental Geosystems* and this fifth edition of the Instructors Resource Manual. The success of the first four editions of this text, the six editions of its companion text *Geosystems*, and of the new *Geosystems Canadian Edition*, is owed to the teachers and students who have embraced this approach to physical geography. The fifth edition is therefore dedicated to you the teachers and the students of this edition. "*To all the students and teachers of Earth, our home planet, and its sustainable future. And to the scientists and researchers, exploring the operations of Earth systems often in the face of political headwinds.*" This edition also begins with a quote from Barbara Kingsolver that reminds us of the crucial role that the natural world plays in our lives. This gifted writer was honored in 2006 by the AAG as the "Honorary Geographer."

    We are now in the first decade of the twenty-first century. Despite the irrelevancy of our calendars in the Cosmos, the century ahead is an excellent opportunity for reflection. We should assess the path taken and the path not yet taken or yet envisioned. Future's thinking and teaching will prove effective strategies to spur student interest in their future directions as Earthlings, supported by the life-sustaining physical, chemical, and biological Earth systems.

    Physical geography provides essential information needed to understand and sustain our planetary journey in this new century. Related to these increasing hazards, Canadian former Minister of the Environment, David Anderson, stated:

Current preoccupation is with terrorism, but in the long term climate change will outweigh terrorism as an issue for the international community. Terrorism will come and go; it has in the past . . . and it's very important. But climate change is going to make some very fundamental changes to human existence on the planet.

Currency is a significant area of *Elemental Geosystems* leadership. *Elemental Geosystems* presents contemporary concepts in the context of physical geography. The book is true to the literature and science from allied disciplines. Examples, among many, include: the moment magnitude scale that updated the Richter scale in 1993; functional landscape models that replace older evolutionary models; the effects of jet contrails on atmospheric and surface energy budgets; a discussion of dynamic ecology models that replace older climax (stable point) succession models; general circulation models used in climate simulations; Dr. Gray's (Colorado State University) latest Atlantic hurricane forecasting model; recent urban heat island studies and remote-sensing image; the November 1998 revision of the Soil Taxonomy. In terms of data, *Elemental Geosystems* is updated through 2005 wherever possible. Examples, among many, include: recent near-record tornado occurrences, the Atlantic hurricane season through 2005, recent earthquake and volcanic activity, record 2005 weather-related damage, record 2005 global land and ocean temperatures.

During the life of this fifth edition of *Elemental Geosystems*, important scientific research and global trends in Earth systems science will unfold. As pointed out on the back cover of the text, climate change and its many impacts are hitting the high latitudes at a faster pace than elsewhere on Earth. To reflect global system linkages with these regions, this edition of *Elemental Geosystems* weaves high-latitude themes through the chapters.

The scientific community is responding with the comprehensive Arctic Climate Impact Assessment (ACIA) Report published in late 2004. This historic scientific effort is detailed in Chapter 17. The International Polar Year (IPY) for science and research in the polar regions is set to begin March 2007 and go through March 2009. The world's scientific community, including geographic scientists, is to focus on the dramatic changes in the high latitudes.

The author of the text and his photographer wife completed five polar expeditions in 2003, 2004, and 2005 gathering materials and photos for this edition. As you study physical geography you will find your textbook responsive to the latest science such as the IPY process.

New to this edition is the "High Latitude Connection" feature, linking chapter material to the polar regions. The world is turning its attention to the high latitudes to better understand present global climate change.

The *Elemental Geosystems* Package

Elemental Geosystems is designed to give you flexibility in presenting your course. The text is comprehensive in that it is true to each scientific discipline from which it draws subject matter. This diversity is a strength of physical geography, yet makes it difficult to cover the entire book in a school term. *Elemental Geosystems* is organized to help you customize your presentation. You should feel free to use the text based on your specialty or emphasis, rearranging parts and chapters as desired. The four-part structure of chapters, systems organization within each chapter, Focus Study and News Report features—all will assist you in sampling some chapters while covering others in greater depth. The following materials are available to assist you—have a great class!

• **Instructor Resource Center CD/DVD (IRC)** (ISBN: 0-13-154761-5) contains everything you need for efficient course preparation. Find all your digital resources in one, well-organized, easy-to-access place. Included on this CD/DVD are:

Figures—JPEGs of all illustrations and, new to this edition, all the photos from the book.

Animations—All the animations ready to use in your classroom presentations as PPT slides or Flash files.

PowerPoint™—Preauthored slides outline the concepts of each chapter with embedded art for use in the classroom or to customize with other items for your own presentation needs.

TestGen—Complete TestGen-EQ software featuring a thorough and revised test bank, with questions and answers.

Electronic files—The complete Instructor's Resource Manual and Test Item File in one easily accessible location.

• **Instructor's Resource Manual, Fifth Edition** (ISBN: 0-13154792-5) by Robert Christopherson and Charlie Thomsen. The Instructor's Resource Manual, intended as a resource for both new and experienced teachers, includes lecture outlines and key terms, additional source materials, teaching tips, complete annotation of chapter review questions, and a list of overhead transparencies.

• **Geosystems Test Item File** (ISBN: 0-13154760-7) by Robert Christopherson and Charlie Thomsen. This collaboration has produced the most extensive and fully revised test item file available in physical geography. This test bank employs *TestGen-EQ* software. *TestGen-EQ* is a computerized test generator that lets you view and edit test bank questions, transfer questions to tests, and print customized formats. Included is the *QuizMaster-EQ* program that lets you administer tests on a computer network, record student scores, and print diagnostic reports. Mac and IBM/DOS computer formats are served. Order the TestGenEQ version of the text bank using ISBN: 0-13-154762-3.

• *Overhead Transparencies* (ISBN: 0-13-154763-1) includes more than 350 illustrations from the text on 300 transparencies—all enlarged for excellent classroom visibility.

• *Applied Physical Geography—Geosystems in the Laboratory, Sixth Edition* (ISBN: 0-13-133093-4) by Robert Christopherson and Charlie Thomsen. The new sixth edition is the result of a careful revision. Twenty-one lab exercises, divided into logical sections, allow flexibility in presentation. Each exercise comes with a list of learning concepts. Our manual is the only one that has its own complete glossary and stereolenses and stereomaps for viewing photo stereopairs in the manual. A complete *Solutions and Answers Manual* is available to teachers (ISBN: 13-133094-2).

• *Elemental Geosystems Student Learning Center*: The Student Learning Center gives you the opportunity to further explore topics presented in the book using the Internet. The site contains numerous review exercises (from which you get immediate feedback), exercises to expand your understanding of physical geography, and resources for further exploration. This web site provides an excellent opportunity from which to start using the Internet for the study of geography. For convenience, all the URLs presented in the text are listed on the web site, along with many additional links. Fifteen Career Link essays feature geographers in a variety of professional fields practicing their spatial analysis craft. Please visit the site at **http://www.prenhall.com/christopherson**.

Access codes for students to fully access this website appear inside the front cover of every text. Contact your local Prentice Hall sales representative (name on file at your bookstore) or contact Prentice Hall directly to receive all these support items.

Geographic Awareness

Undergraduate physical geography is an introduction to geography and as such is an opportunity to teach general education students and perhaps interest them in our discipline—a potential harvest of teaching, academic, and practicing geographers for the future. At our American River College, classes are filled with freshman and sophomore students completing their core curriculum courses before transferring to upper-division work. Physical geography is an introduction, and perhaps their only exposure, to science and scientific methodology. Because of the nature of physical geography, a diverse and dynamic approach seems most appropriate.

Popular news sources have carried many articles concerning the geographic illiteracy of students and the general public. And, strangely enough, the subject of global geographic education became politicized in the 1980s with former Secretary of Education Bennett actually warning the nation against eliminating such illiteracy! Unbelievably, educators were told by this former official that students would "suspend judgment and get wishy-washy" if they knew their geography and history better! A clearly ridiculous statement on the face of it!

We suggest that part of a lecture period be devoted to a discussion of geography and the importance of knowing the "where" of physical and cultural things—the spatial component is involved in more than 80% of business decisions according to one study. This is why the text author begins Chapter 1 with a section defining the geographic approach and our scientific discipline. Featured are the five themes authored by the Association of American Geographers and the National Council for Geographic Education. The course can be dedicated to placing each student consciously onto the world stage and in their campus and community—at least in terms of physical systems.

In the Preface of the *Applied Physical Geography: Geosystems in the Laboratory*, sixth edition, manual students are asked to complete their "Geography I. D." The one-page assignment is a personal application of the five themes to thoroughly locate and place the students in the environment. We include this one-page geography lesson as the last page of this introduction. Let us know what you think.

The Association of American Geographers and the National Council for Geographic Education are both active in programs that improve public geographic literacy and upgrade geographic education. The National Geographic Society, assisted by many geographers and educators nationwide, has set up over 60 Geographic Alliances to promote geographic education. You may want to consult your local alliance and acquaint your students with their programs. These alliances produce newsletters and information useful for teaching. Contact the Geography Education Program, Educational Media Division, at **http://www.nationalgeographic.com/ education/teacher_community/index.html** for more information about state alliances, availability of materials for geographic education, and to sign up for their excellent monthly e-newsletter.

By federal mandate, we now have an official "Geography Awareness Week" each November.

Also, the National Geographic Society conducts a National Geography Bee beginning at the state level each April and ending with national finals held in Washington, D.C., in late May hosted by Alex Trebek from *Jeopardy!* Every two years they sponsor the National Geographic World Championship (see Figure 17.9 for a photo of a recent championship). Geography is now specified as one of five key core curricula for national emphasis, along with history, mathematics, science, and English. As a discipline, geography has gone through a strong resurgence over the past two decades.

Geography is a discipline in growing demand as humans place ever-increasing stress on the Earth-human relationship. Career opportunities involving geographic information systems (GIS) are flourishing in many fields. The demand for environmental analysis and assessment increases each year. We feel it important to keep these realities on the front burner during the term, to constantly bring this relevance into class.

Format for This Manual

In this resource manual, each text chapter is supported from several perspectives. You will find each chapter of this manual organized in the following manner.

Overview: A brief introductory statement summarizing the content of the chapter and our overall goals for the text presentation.

Outline Headings and Key Terms: A listing of all orders of headings in the chapters and key terms and concepts presented, with a convenient check box to keep track of progress.

Key Learning Concepts: A list of the key learning concepts that appear at the beginning of the chapter and that are used to organize the Summary and Review section.

Annotated Chapter Review Questions: The review questions at the end of each chapter are specifically answered with wording from the text. These are grouped under each of the key learning concepts, as they are in the chapter Summary.

Overhead Transparencies List: As an adopter of *Elemental Geosystems*, you received a packet of 300 high-quality, full-color overhead transparencies of figures from the text. The overheads selected for each chapter are listed.

Critical Thinking Essentials

"Critical thinking" encompasses the deliberate and conscious control of the thinking process by the thinker. When the essentials of critical thinking are integrated into teaching the classroom process leaps to exciting levels.

Note the new "Critical Thinking" sections at the end of each chapter in *Elemental Geosystems*, fifth edition.

Do not confuse critical thinking practices and content; rather it should be thought of as the organization and conceptualization of content and the tactics and strategies for its delivery. Instruction should deliver content in a way that helps the student learn to reason and assists them to see the application of the material to their lives. Learning is defined as a change in behavior in the learner, not merely knowledge acquisition. Critical thinking effectively leads to such change and improvement in the learner. Such thinking assesses and improves itself continually.

A government report (Department of Labor, 1991) summarized the ideals of such a learning paradigm: development of thinking skills, assessment and testing integral to teaching, cooperative problem solving, information in the context of real problems, learner-centered and teacher-directed classrooms, and all students learn to think.

We should give the students the tools to ask: What do I want to accomplish? How can I measure if I am accomplishing this goal? What are my questions? What do I need to answer them? What other points of view are available to me? Why did I select the point of view I now have? Can I justify my answers or have I missed something or taken something for granted? How clearly am I expressing through speech and writing my ideas and findings? Can I still identify my path and direction of travel to my goal? Where am I at right now? How does this position relate to my beginning goal statement?

Critical thinking requires that we as teachers set out our goals for physical geography at the outset of the class: purpose of the class, its importance to our lives, the method of assessment and why this method is part of the purpose, a statement of the questions we want to answer, and the level of emotional, mental, and social energy the course will tap. This deliberate structure will signal a lot to the student and put the term on a higher academic plane—*one that involves the individual in the process.*

To assist, we have included in this manual sections on "Teaching Essentials" and "The Teacher: Three Aspects." Also, we include with each chapter of this *Instructor's Resource Manual* and the *Student Study Guide* a set of key learning concepts (behavioral objectives) to set the table for each chapter. The rigorous use of a three-ordered outline system for the textbook facilitates a critical thinking approach to learning.

For more on critical thinking, see Richard Paul, *Critical Thinking—How to Prepare Students for a Rapidly Changing World*, Foundation for Critical Thinking, 1993; Linda Elder and Richard Paul, "Critical Thinking: Why We Must Transform Our Teaching," *Journal of Developmental Education* 18, no. 1 (Fall 1994): 34–35; and the publication *Educational Vision—The Magazine for Critical Thinking*, published four times a year by The Foundation for Critical Thinking, P.O. Box 220, Dillon Beach, CA. 94929, 1-707-878-9100. See **http://www.criticalthinking.org**.

The Internet and *Elemental Geosystems*

The fifth edition of *Elemental Geosystems* presents more than 200 URLs (Internet addresses) right in the text. Students intrigued by a subject are able to go on the Net and find the latest information using these many links between the text and the Internet. This innovation should enliven your teaching experience with the text.

We now live on a planet served by the Internet and its World Wide Web, a resource that weaves threads of information from around the globe into a vast fabric. The fact that we now have Internet access into almost all the compartments aboard Spaceship Earth is clearly evident in *Elemental Geosystems*. Such ready availability of worldwide information allowed the author to illustrate content with a fascinating array of up-to-date examples and to further verify accuracy.

The Elemental Geosystems Student Learning Center gives your students on-line review, critical thinking exercises, tests that are graded on line, and opportunities to delve deeper into subjects out on the Net through our exclusive "Destinations" link feature that ties the chapter directly to numerous URLs. There are follow-up answers to specific items in the text, and annual precipitation, temperature, and potential evapotranspiration for 30 cities. Again, the URL (Net address) is:
http://www.prenhall.com/christopherson

A Tour of *Elemental Geosystems*

Earth's physical geography systems are complex, interwoven threads of energy, air, water, weather, climates, landforms, soils, plants, animals, and the physical Earth itself. *Geosystems* is thoroughly up-to-date, containing the latest information about the status of Earth's physical systems as viewed through the *spatial analysis* approach. *Elemental Geosystems* is an introductory text in which relevant human-environmental themes are integrated with the principal core topics usually taught in a course of physical geography. The text is conveniently organized into four parts that logically group chapters with related content. After completion of Chapter 1, "Foundations of Geography," these four parts may be taught in the integrated text sequence or in any order convenient to the instructor.

Elemental Geosystems is organized around process systems and is essentially non-mathematical, with text material organized in the same direction as the flow of energy and matter, or arranged in a manner consistent with time and the flow of events. Particular attention was invested in the rewriting and revision of *Elemental Geosystems* to achieve readability and clarity appropriate to the student in university- and college-level courses. The student will benefit from thorough review questions, a comprehensive glossary, and full-color high-production values as evidenced by more than 500 figures, many presented as compound—mixing photos, art, and cartography. *Elemental Geosystems* contains hundreds of photographs, many by the author and more than 320 by his wife, and 90 remote-sensing images from an array of orbital platforms.

Physical geography is an important Earth system science, integrating many disciplines into a complete portrait of Earth. *Elemental Geosystems* analyzes the worldwide impact of environmental events, synthesizing many physical factors into a complete picture of Earth system operations. A good example is the 1991 eruption of Mount Pinatubo in the Philippines. The global implications of this major event (one of the largest eruptions the twentieth century) are woven through seven chapters of the book (see the new Figure 1.6 for a summary). Our update on global climate change and its related potential effects is part of the fabric in six chapters. These content threads, among many, weave together the variety of interesting and diverse topics crucial to a thorough understanding of physical geography.

You will notice many new and re-rendered figures and revised text sections in the fifth edition. These changes are noted in the chapters that follow in this guide. The fifth edition is a thorough revision and is shorter than the companion text. Yet you will notice many new photographs, art, and maps. This was accomplished through careful rewriting of text, assembling more compound figures, focus on core content, and an increase in trim size. This increase in page size allows the visual aspect of the book to be fully developed for the first time.

As in the past, the text author has brought content up to date with the latest information, data, and research, being faithful to the exciting scientific breakthroughs of this time and true to the disciplines from which we draw content for spatial treatment. Wherever possible the most recent data are used.

Every chapter begins with a list of "Key Learning Concepts" that tell the students what they should learn from the chapter. At the end of the chapter is a "Summary and Review" section that is organized using these initial Key Learning Concepts. Under each concept are a narrative defining all the key terms presented in the chapter, a list of key terms with page numbers, and a series of review questions.

Each chapter ends with a "Critical Thinking" section of several thought/action items to stimulate the student further. These ask the student to actually do some tasks, or apply chapter content to a problem, or make personal observations, or work with Internet resources.

Nineteen "Focus Study" and 35 "News Reports," many with figures, highlight key topics related to text material. The emphasis varies among the Focus Studies from traditional to topical to environmental to contemporary Earth systems science subjects. Some present specific scientific tools as in focus studies on the scientific method and temperature concepts. The News Report feature boxes cover timely topics such as GPS, Careers in GIS, the story of a man who jumped out of a capsule at 34 km (21 mi) altitude and fell and parachuted to research the atmosphere, wind and solar power applications, forecasting Atlantic hurricanes, High Plains Aquifer overdraft with the latest maps, water in the Middle East, the 1996 and 2003 experiments to scour the bed of the Colorado River in the Grand Canyon, and the Greenland and Antarctic ice cores and their records.

Elemental Geosystems presents the latest conclusions from the Intergovernmental Panel on Climate Change (IPCC), the seasonal changes in spring and fall timing due to global warming, a new scientific consensus report on stratospheric ozone disruption, the results of the 1997 Kyoto Conference to reduce carbon emissions, the Russian ratification and subsequent adoption of the Kyoto Protocol and Rulebook as international law, the Council of the Parties meetings through 2005 (*COP-11*), an analysis of the benefits of the Clean Air Act, and the latest changes in the Antarctic ice shelves and information from Greenland and Antarctic ice cores.

The following are examples of coverage: sections on future temperature patterns; changing weather and climate trends; reduction in the volume of glacial ice and the recent disintegration of Antarctic ice shelves; rising global sea level; changes in atmospheric and surface energy budgets; variation in expected weather patterns; trends in storms, tornadoes, and hurricane activity; dynamic changes in natural habitats and ecosystems, changes in soil moisture, and losses in biodiversity and species extinctions—all data through 2005 wherever

possible. This timeliness in the book is unique in our market.

Let's begin this brief tour by examining the book itself. The dramatic front cover is a photograph of the highest sea stack in the British Isles, *Stac an Armin*, in the St. Kilda Archipelago, 165 km (103 mi) west of northern Scotland in the Atlantic Ocean made by Bobbé Christopherson. This beautiful image captures the four spheres that make up Earth: the atmosphere, hydrosphere, lithosphere, and biosphere.

Geography and physical geography and Earth systems science are thoroughly defined in Chapter 1 and applied throughout the text (see first five bold-faced/glossary key terms). Figure 1.1 demonstrates the five themes of geography. Figures 1.3, 1.4, 1.6 demonstrate systems theory and how systems organize the textbook. A world map showing a systems analysis of the Mount Pinatubo eruption in Figure 1.5 draws together later chapters as an example of how process-systems in a real-time Earth event work.

The Scientific Method is discussed and applied in Focus Study 1.1, with a flow chart showing how it operates. This establishes right at the beginning that physical geography is an essential Earth systems science.

All maps feature complete geographic content, latitude, longitude, and scale (see any world map presentation in the text for example of quality cartography). See the innovative map comparing strike-slip faults in Turkey (the New Anatolian fault system) and California (the San Andreas fault system) in Figure 9.13.

Space does not permit a listing of all the changes to *Elemental Geosystems*, fifth edition; however, here are a few comments. Note in the Table of Contents that "Global Climate Systems" is now the culminating Chapter 7 in Part 2 and that "Water Resources" is in its logical position as Chapter 6. To begin, Chapter 1 treats the engineering feat of moving the Cape Hatteras Lighthouse 884 m (2900 ft) inland, in terms of its changed latitude and longitude grid coordinates. A compound map, satellite image, and photos cover the same area in Figure 13.13.

Examine all 17 chapter-opening and 4 part-opening photos and read their captions—all except one are new. These are content-specific and begin the learning process from the start of the chapter. Throughout the text there are hundreds of new photos to illustrate concepts, many integrated in compound figures with art and maps. For example, in Chapter 14, "Glacial and Periglacial Landscapes," alone, there are 34 new photos and 3 new satellite images.

The year 2005 broke many records climatologically. This warmest year in the past several hundred thousand years for meteorological stations is reflected in the global average temperature graph in Figure 7.26a and accompanying text. The lowest extent of Arctic Ocean pack ice in the record occurred during 2005, featured in News Report 3.1 and Chapter 14 in Figure 14.29b and text. Figure 4.3 reflects the new record low pressure for North America set by Hurricane Wilma in 2005, with hurricanes Gilbert (1998, the previous record low), Rita, and Katrina (both 2005) listed for comparison.

New to this edition is the "High Latitude Connection" feature, linking chapter material to the polar regions. The world is turning its attention to the high latitudes to better understand present global climate change. These features include:
1.1: Meltponds as Positive Feedback
3.1: Overview of Trends in the Polar Regions
7.1: Climate Change in the Polar Regions
8.1: Isostatic Rebound in Alaska
13.1: A Rebounding Shoreline and Coastal Features
14.1: Climate Change Impacts Arctic Ice Shelf
17.1: Report from Reykjavík—Arctic Climate Impact Assessment

Chapter 14 covers the ACIA Symposium in Iceland in November 2004. The report that emerged from this scientific process is summarized in this book. There are two additional related photo galleries on the *Student Animation CD*. Also on the CD, in a special "High Latitude Connection" feature, five new video clips made by the author in Arctic and Antarctic regions show students firsthand the look, feel, and sounds of these changing land and seascapes. New polar region temperature maps are in Figures 3.24 and 3.26. In addition, the book includes new coverage of regional circulations of the Arctic, North Atlantic, and Pacific Decadal oscillations in Chapter 4, as well as the latest science reporting on the thermohaline circulation in the deep ocean. The impact of surface ocean freshening (reduced salinity) is apparently causing nearly a 30% slowing of this circulation. Also in Chapter 4 is a new News Report 4.1 on the MANTRA hot-air balloon that inadvertently demonstrated jet stream flows.

Chapter 5 includes an exclusive discussion of derechos—straight-line (non-tornado) winds that annually cause so much damage. Coverage of the May 2003 tornado outbreak is in News Report 5.4 with a new map, along with all updated tornado statistics. Figure 5.40 presents a world tropical cyclone map with two added satellite-image insets of bizarre, first-ever hurricanes that hit Spain and Brazil. The 2005 Atlantic tropical storm season, featured on a new map in Fig. 5.41a, set many records, including:
• 27 named storms (average is 10)
• 14 hurricanes (average is 5)

• 7 intense hurricanes, category 3 or higher (average is 2)
At this writing it appears that the National Hurricane Center has reevaluated Tropical Storm Cindy (7/3/06 to 7/6/06) and upgraded it to a category 1 hurricane. Therefore, you might see the final total for 2005 listed at 15 hurricanes.

Chapter 5 presents analysis and explanation of such violent weather and features as a new map and satellite images.

Chapter 8 features a new author-drawn map in Figure 8.3 that presents Earth's interior cross-section profile with Earth's center placed at Anchorage, Alaska, and the surface crust at Fort Lauderdale, Florida. Also, the chapter includes new photo examples of rocks and landscape features throughout.

In Chapter 9, updated discussions cover the October 2005 Pakistan earthquake that killed 83,000; and the Sumatra earthquake and resulting tsunami that took more than 170,000 lives is explained in detail in News Report 13.2 and with two satellite images showing the spreading tsunami at 2 hours and 7 hours after the quake.

In Chapter 10, see News Report 10.1 on the weathering of bridge stone in Central Park, New York City, and new photos in Figures 10.6, 10.9, and 10.20 showing sheeting (exfoliation) and talus slopes from Spitsbergen Island and Nordauslandet Island, Duve Fjord, in the Arctic Ocean. In Figure 10.10, see the weathering of a cathedral built in A.D. 1137 in the Orkney Islands, Scotland. And Figure 10.13 features a new, revised world map of karst distribution.

News Report 11.3, "What Once Was Bayou Lafourche," is a poignant story of a hurricane levee compound that is slightly west of where Hurricane Katrina made landfall—updated to reflect the aftermath of 2005 events. Expanded coverage of the Mississippi Delta appears in Figure 11.26a–h and Focus Study 11.1.4. And new photos of Bonnet Carré spillway in New Orleans and river flood avoidance structures are in Focus Study 11.1.

In Chapter 12, Focus Study 12.1 has the latest information on the troubled Colorado River system, a subject discussed in all my books since the first edition. This continuing focus study accurately defined the present crisis in advance and has followed it through to this new predicament of record-low flows since 2002. Note the tie-in to the Chapter 6 opening photo comparison showing Lake Mead levels at Hoover Dam.

Chapter 13 illustrates the impact on coastal regions and barrier islands, such as the loss of the Chandeleur Islands in the Gulf of Mexico (Figure 13.14).

Focus Study 14.1 has expanded coverage of Dome C, Antarctica, and the ice-core analysis that now stretches the climate record back to 650,000 years ago, confirming present levels of greenhouse gases in the atmosphere as the highest on record. News Report 14.3 is new coverage of the increase in meltponds in Greenland and the climatic causes and implications.

Coverage of Histosols in the Soil Taxonomy is expanded with a chapter-opening photo of cut and dried peat in the Orkney Islands, north of Scotland, among four new soil photos in the chapter.

In Chapter 16, a new section, "Biological Evolution Delivers Biodiversity," describes the theory of evolution and its spatial implications for producing the biodiversity on Earth. See the new News Report 16.1 documenting the "Chinstrap Penguins and Their Short, Busy Summer." And shading has been added to Figure 16.10a to demonstrate the probable spread and path of the "muck" plume of pollution from the Hurricane Katrina discharge along the Gulf Coast as it spreads northward in the Atlantic Ocean—among many new items in this chapter.

Chapter 17 presents a chapter-opening composite of five photos of the damage from Hurricane Katrina along the Gulf Coast; Figure 17.7 contains five more damage scenes and an inset showing water-level stains on houses; Figure 17.8 offers two before and after comparisons. A new heading in Chapter 17, "2005—A Year for the Record Books—Lessons Learned?" summarizes these record breaking events. Newly recast for this edition, Chapter 17 ends the text with a unique capstone chapter, "Earth and the Human Denominator," that summarizes human-Earth interactions and presents a guiding philosophy for the coming century by answering the question, "Who speaks for Earth?" and the most profound issue of our time, *Earth stewardship*. The final chapter brings together physical geography and the entire text in a thoughtful list of "12 Paradigms for the New Millennium." From Chapter 17 of *Elemental Geosystems*:

Twelve Paradigms for the 21st Century
1. Population increases in the less-developed countries.
2. Planetary impact per person (on the biosphere and resources; $I = P \cdot A \cdot T$).
3. Feeding the world's population.
4. Global and national disparities of wealth and resource allocation.
5. Status of women and children (health, welfare, rights).
6. Global climate change (temperatures, sea level, weather and climate patterns, disease, and diversity).
7. Energy supplies and energy demands; renewables and demand management.
8. Loss of biodiversity (habitats, genetic wealth, and species richness).
9. Pollution of air, surface water (quality and quantity), groundwater, oceans, and land.
10. The persistence of wilderness (biosphere reserves and biodiversity hot spots).
11. Globalization versus cultural diversity.
12. Conflict resolution.

Appendix A covers mapping and topographic maps and presents a section on the maps used in this text.

Appendix B presents the Canadian System of Soil Classification (CSSC) with a regional color map of Canadian soils.

Köppen climate classification information has been placed into Appendix C, along with a map showing the Köppen zones along with the causal factors like ocean currents and air mass source regions.

The text and all figures use SI and other modern systems of measurement equivalencies and English unit conversions. An expanded set of measurement conversions is conveniently presented in an easy to use two-page arrangement in Appendix D. Earth radiation balance maps and graphs, as in all figures and text, appear in SI and metric units with English equivalents.

For more information about metric conversion materials and programs contact: the Metric Program Office, National Institute of Standards and Technology, U.S. Department of Commerce, Gaithersburg, MD 20899. A useful report was prepared by Barry N. Taylor, ed., *Interpretation of the SI for the United States and Metric Conversion Policy for Federal Agencies*, NIST Special Publication 814, NIST, Department of Commerce, October 1991. And nongovernmental organizations: the U.S. Metric Association, 818-368-7443; HRD Press, 1-800-822-2801; Marlin Industrial Division, 1-800-344-5901. The conversion law is Public Law 100-418, 100th Congress, H.R. 4848, 8/23/88, the "Metric Conversion Act" that amended the original 12/23/75 act.

On the inside back cover of the text you find the new *Elemental Geosystems fifth edition Student Animation CD*. This exciting CD contains 67 new animations, 18 satellite loops, 5 High Latitude Connection videos, 3 narrated photo galleries, and instantly graded self-tests.

Teaching Essentials

We do not know if there is a correct way to teach or if any individual has discovered the "right" approach. This is part of teaching's mystery and intrigue. The variation of practices among teachers is as subjective and variable as the individuals themselves. As more experience and maturity is gained in the individual personality, so too does the definition of the teaching craft evolve within that individual. The teacher is constantly redefining the inner "ultimate potential teacher" concept. So to pin down specific rules for teaching is impossible, although some generalizations are appropriate.

It is time that we all accept the status of teaching as a true profession and that practitioners are true professionals. No matter how hard we stress academic geography and published research, the flow of new geographers into the discipline and the enrollments so vital to funding and administrative internecine campus battles are both keyed to our ability to attract people to our classes. This enrollment in turn generates interest from the institution. Think for a moment of the master teacher who started your inner fire.

Dr. Susan Hanson, past-president of the Association of American Geographers, summarized the importance of teaching and teaching methodologies in her column in the January 1991, *AAG Newsletter* (Vol. 26, No. 1, p. 1):

> ...an important message for academic geographers: good teaching is essential to the renewal and vitality of our discipline, and now is the time to capture students' imagination and idealism by conveying the conviction that geographic insights can empower people to make a difference....With students now clearly wanting to contribute to solving problems like urban poverty, environmental pollution, uneven development, land degradation, and resource depletion, we have the opportunity to demonstrate the strengths of the geographic tradition in tackling problems. The key is effective teaching.
>
> It is time to see that the strength of our discipline lies not only in pursuing new knowledge about the world but also in infecting others with an eagerness to pursue new geographic knowledge on their own. It is time to openly acknowledge and cherish and learn from the good teachers among us. It is time to reflect closely upon, and to share with colleagues, what happens on those days when our teaching "clicks" and we leave the classroom with that glow that comes from knowing students have caught the fire of our enthusiasm. It is time to think anew about how people learn, and about how we might tune our teaching to accommodate the diversity of learning styles in our classrooms. It is time to take advantage of new technologies that facilitate learning.

Teaching Media. The student of today is visually motivated to a high degree and therefore responsive to media. If you are fortunate enough to have a dedicated geography classroom, then a plan can be worked out among instructors for making it part of the media for teaching. Possible media resources may include the classroom walls and ceiling (for display of maps, charts, satellite images, an events bulletin board, and posters), PowerPoint presentation capability, overhead projector, slide projector with remote control at the screen, video monitor and playback with remote, computer projector system, cassette tape player, shortwave radio, chalkboard, and a teaching "toolbox."

The "voice in the shadows" and "this is a..." type of presentation is dead in the dark. Make sure your laptop or computer is located so you can stand by the screen. Consider using a radio-frequency remote mouse so you can move around the classroom, rather than being tethered to the computer. Use slides the same way you would tutor a student with a figure in the text, pointing things out with your hands in the picture. Even though you cannot see the eyes of the students in the classroom, look out into the projector beam as if you can see their eyes, much as stage actors learn to make connection with their audience.

• **Visualization software.** We have placed this first due to its potential impact. The advances in computer visualization software are staggering. There are currently two free software products that could radically transform your teaching. They are both very similar, yet they have different capabilities.

The first, Google Earth, is available for both PC and Mac computers. It is a free download from **http://earth.google.com**. You install it on your computer and it streams data across the IInternet. If you have a high-speed Internet connection and a digital projector in your classroom there is no reason not to use it. You can specify the vertical exaggeration of the landscape, the tilt and rotation, zoom in and out, create bookmarks, and fly all over the world. We have used this in the classroom to examine river meander bends, glacial moraines, the trace of the San Andreas fault, deranged drainage, and many other features. It is like having a private plane at your disposal. If you lack a high-speed connection in your classroom the program allows you to save screen images for later display.

The second program is World Wind, and is available for download from NASA at **http://worldwind.arc.nasa.gov**. Unfortunately for all of you Mac users, this is currently available only for PCs. The visualization options are the same with regards to zooming, tilting, rotating, and specifying vertical exaggeration, but World Wind allows you to add overlays of different data. Did you want to see a month by month progression of snowpack? Add the Blue Marble layer. Before and after flooding due to Hurricane Katrina? Add the MODIS theme. Sea surface temperatures during an El Niño? Turn on the Scientific Visualization layer. There are *LandSat 7* overlays with different color schemes to highlight features such as vegetation patterns or geologic formations. You can display weather data from the Naval Research Laboratory—I showed my class all the storms in the Pacific Ocean from January 1, 2006, to March 29, 2006, as an animation draped on the globe. You can also access MODIS imagery and animations such as the progression of Hurricane Katrina or the deforestation of Rondônia as shown in Figure 16.27. The abilities of this tool will hopefully continue to grow as the availability of imagery increases.

• **Computer and Internet.** This is the single greatest change in instructional technology in the past several decades. When used wisely a computer, the Internet, and a digital projector can be an extremely powerful combination for in-class presentations. For those without a live hookup in the classroom, most of the following examples can be adapted for use with either a digital projector and computer or with an overhead projector.

Be sure to contact your Prentice Hall sales representative and ask about the availability of the **Instructor Resource Center CD/DVD**, (0-13-154761-5), which contains most of the figures and photographs from the text, along with PowerPoint slides and Shockwave Flash animations, showing crucial concepts from the text.

It is difficult to think of a chapter that does not lend itself to examples from the Internet. In fact, many of the suggestions from the previous edition of this guide (shortwave radio, cassette recorder, slide projector, overhead projector, video monitor) can be improved upon with a computer/digital projector combination. In Chapter 1 the official U.S. time site (*http://www.time.gov*) has a live simulation of the illumination of Earth. Once you set your time zone you can view the terminator line between day and night. Over the semester you can check back and your students will see how the terminator is a vertical line at the equinoxes, while at the solstices one of the poles will be constantly dark. For Chapter 5, you can access real-time weather data and images. Students can follow the process of cyclogenesis with a storm that is moving overhead! Many weather sites provide satellite images with front overlays, often with an animated version. Your students can track the relationship between temperature and humidity from the beginning of class to the end. Students are amazed by how quickly the numbers roll by on the Census Department's Population Clock, **http://www.census.gov/ipc/www/clock.html.** For added impact, you can visit the site and have them record the world and U.S. population when you begin Chapter 17, and then visit it again when you are done with the chapter. They (and perhaps you) will be surprised by the increases.

At the beginning of the semester in 1997, less than 10% of the students had actually been online and even less had sent and received E-mail. By the 1999 school year, a classroom survey at the beginning of the semester found the Internet was used by over 90% of students—a remarkable increase in market penetration! Of course, today the use level is right near 100%. The student assessment was overwhelmingly positive, as they immediately began using the resource for all their classes. By the way, the *Elemental Geosystems* Student Learning Center allows students to E-mail work directly to you.

The *Elemental Geosystems* interactive Student Learning Center works well in classroom demonstrations—illustrated critical thinking items, illustrated short essays, destination links, and more.

Many of the other uses of the computer and Internet and projector combination are replacements for existing technologies. Streaming media and mp3 podcasts can replace cassette tapes and shortwave radio broadcasts. DVDs and other digital video formats are replacing VCRs. There is a wealth of videos available on the Internet, and the days of slides are numbered with Kodak's announcement that they will no longer manufacture slide projectors. For those of you with extensive slide collections this may present a dilemma. At American River College the Instructional Media Center will perform bulk conversions of slides to jpgs, as well as video and audio tapes to digital formats.

One last recommendation involving the Internet is to use an RSS reader to scan headlines and automatically deliver news stories. RSS stands for Rich Simple Syndication. The RSS program can be a standalone program or incorporated into a web browser like Firefox. The RSS program allows you to "subscribe" to a web site and periodically view what is new, thus saving you from having to read through the entire site. These subscriptions are almost always free, even for sites supported by paid print subscriptions, such as the journals Science and Nature. For more information see the following: **http://en.wikipedia.org/wiki/RSS_(protocol)**

• **Classroom.** Posters, photographs, images, and maps that relate to topics under discussion can be placed in different parts of the room so that you can leave the front and teach from other locations in the room. For example, a Mount St. Helens poster, map projections chart, image of Hurricane Mitch, relative age of the ocean crust and ocean floor maps, weather map symbols, and world physiographic map could be mounted around the room. With a poster of the Andromeda Galaxy in the back of the room you can walk the 2 million light years to it during that discussion. The National Geographic Society's "The World" mural map (120" x 72") is reasonably priced and suitable for permanent installation. If your materials must be transported, it might be convenient to place those items that are not on rollers on large pieces of cardboard. (Appliance and automotive dealers throw away such cardboard. Various acetate coverings are available from suppliers.)

It is helpful to have two screens placed at angles in each front corner—one for overheads and the other for slides. In this way both screens and the whiteboard can be used simultaneously. The video monitor is installed in the front-center of the room above the whiteboard, with the VCR/DVD installed in a modified cabinet drawer on the side of the room. In other rooms, a mobile cart is used, a situation probably similar to that on most campuses. The ceiling-mounted computer projector uses one of the screens for projection.

We recommend transitioning from chalkboards to whiteboards if you still use this medium. Your presentations will gain visual impact, and your electronic equipment will be safe from chalk dust. Felt-tip dry-erase markers come in an odorless form and a box lasts about a semester. The Sanford Expo® brand features eight colors, suitable to portray components of most physical systems. Students express appreciation for the use of color for board work. You might want to keep your own set of whiteboard pens (and possibly an eraser) and do not depend on pens left in the classroom.

If you must use chalk here is a suggestion. Buy a high-quality colored chalk, such as HygaColor, that is neither too hard nor too soft and easily erasable.

Color, either pens or chalk, can be suggestive of the subject at hand and can prove valuable in giving the student some visual variation. Our texts are in color, reality is in color, so why not our pens and chalk!

Light levels in the classroom are very important. You will benefit from being able to vary the illumination from having all the lights on to having all the lights off. The option of total darkness may depend on your schools emergency exit illumination policy, but it is useful for displaying dark overheads or images. A note-taking light fixture in the back of the room gives enough light for students to take notes, but media can be shown in the darkened front portion of the room. This back light, and indeed all your lights, can run from switches in the front of the room so you do not have to run back and forth to adjust the light levels.

• **Overhead projector**. With the availability of thermal films and copy machines, it is possible to make overheads from any source. Students take note when items appear on the overhead carried in that morning's paper. As dictated by room size, the overhead can be used in concert with the chalkboard. We do not recommend its exclusive use in rooms with under 100 chairs because an inherent visual disconnection between the teacher and the student is associated with its chronic use. In larger rooms, however, nearly continuous overhead use is necessary. You have available over 300 figures from the text for your overhead transparency set, all appearing enlarged with enlarged labels.

• **Slides.** Slides are an essential resource in physical geography because we study the physical elements of Earth's environment. Pictures from many sources, including personal photos, bring the world into the classroom, either directly shown or scanned and used with a computer projector. Any camera with a 50 mm-macro lens can be used to take slides for class from various sources. If a copy stand is not available, then simply use your digital camera or daylight film and set up outdoors with diffused natural lighting—the text author has made many slides on a shaded part of his patio. Or using tungsten film, you may make slides on the dining room table or your desk with two 75-watt floodlights in simple reflectors. Use slower films (ASA-50 or -64, daylight, or tungsten) for finer grain and clarity. Of course, digital cameras and Photoshop simplify their considerations. (It is important to check copyright laws with all copied materials. Be sure the slides you make are not used for profit and that they are exclusively for classroom purposes only.)

• **Video monitor and playback**. Any media can be abused through overuse, and this includes science shows especially when they are used with the teacher out of the picture. We believe that the use of video in a manner similar to the use of slides, like single-concept pictures in motion, is preferable.

Simple editing to produce succinct pieces is easy with two VCRs placed side by side, using an SP mode to produce a better second-generation image. The VCR remote allows you to freeze-frame and reverse for repeated showing, especially useful in creating weather loops. An excellent use of this is for coverage of environmental events such as earthquakes, floods, hurricanes, and so forth. Public domain NASA videos and live events are shown on many local cable channels. You can have shuttle orbital images of Earth running as the students come into class to set the tone for lecture. Check with your local Public Broadcasting System station for more information on availability of videography or with your local cable company for NASA coverage. Your campus probably has a site license for many of these shows. Additionally, C-SPAN I and II allow you to tape key congressional debates and votes, pertinent lectures, hearings, and national conferences that deal with environmental topics associated with physical geography. These video segments can work effectively to augment lectures and to bring a point home.

The *NASA Select* channel provides real-time coverage of each of the shuttle missions as they unfold. In October 1994 one mission transmitted visible and radar images of an eruption on the Kamchatka Peninsula, Russia, as it was occurring! Figure 1.22 shows a photo and a radar image from this mission, both made by astronaut Dr. Thomas Jones, who made the picture of Mount Everest in News Report 9.1, Figure 9.1.2. These events can be taped and used in class. The JPL *Pathfinder* Mars mission press conferences were run on NASA Select and C-SPAN, as an example.

If you have access to a video camera, you may want to tape some lecture segments at a location in the field. The outdoors is certainly usable in class, for example, to incorporate interviews with park rangers and technicians and personal on-camera explanations. The author has taken a camcorder to AAG and other professional meetings to show students details of the meeting when he returned from the trip. This use brings a sense of immediacy to the classroom. On several occasions the author pre-taped a lecture in a natural setting and had it shown when he was away on a travel leave to a professional meeting, thus not losing a class session.

Recently, the text author videotaped volcanic activity in Hawai'i from a helicopter. This taped coverage really went over well in class in illustrating aspects of Chapter 9!

• **Cassette tape player.** National Public Radio news shows have featured segments on geographic education and physical geography suitable for classroom use. When introducing certain topics, the use of music or the sound of waves, rain, waterfalls, and thunder, etc. to accompany slides is a nice feature that varies the student's sensory experience, possibly evoking a feeling about the meaning or beauty of a subject.

• **Shortwave radio.** The author has used a shortwave radio to play Coordinated Universal Time (UTC), solar forecasts, and related foreign broadcasts for classes. Such a radio can also help in the discussion about the role of the ionosphere in wave propagation.

• **Teaching toolbox.** To avoid the before-class scramble for felt-tip pens, laser pointer, chalk, overhead pens, stapler, tissues, paper clips, throat lozenges, it's useful to have a small reinforced box (shoe-box size) and simply take this "toolbox" with you lecture. No matter what shape the room was left by the last instructor or janitor, the toolbox has you ready to go at the beginning of class.

• **Group Activities.** Breaking up the class into groups for discussion and resolution of some issues works better than class discussion and should be used occasionally. Discussions are augmented by using "reaction bubble" diagrams on the board where you record whatever the students say or have determined. In short order, the board is filled with a vast array of interconnected concepts. The author developed the "Twelve Paradigms for the 21st Century" mentioned at the beginning of this Introduction and in Chapter 17 of *Elemental Geosystems*, third edition, following numerous group discussions in the classroom.

Grading and Course Structure Suggestions. All of us develop our own grading system and course structure and then we generally never discuss our system with anyone. It seems as though this becomes the big, sacred mystery! The following in no way is meant as a definitive statement, but rather as a few suggestions—for we are sure we all have good ideas.

• **Course syllabus.** This should include a clearly stated presentation of catalog description, grading policies, goals, test make-up policy, office location, office hours, mail box location, E-mail address, and office phone number. To facilitate *critical thinking*, we must have a clearly stated syllabus at the outset of our courses. Voice mail on the office phone is a great help, increasing student contact easily by a third. E-mail is increasing student contact as well. Students can E-mail you with questions about class, the text, grades, and asking for help with topics for other classes. Of course, getting them to visit your office is the best contact.

• **Grading.** One approach is to structure the class assignments and tests based on 1000 points. This is made up of three equally-weighted exams. The remaining points are filled with various items. Before the *Student Study Guide* was available, the author used four exercises, representing 25 percent of the grade. Relative to testing, please see the separate test bank that we prepared to accompany *Elemental Geosystems*

Exercises may take many forms but can include a library exercise, a weather observation exercise, climographs and a sample station exercise, an article analysis exercise, many versions of mapping exercises, an Internet Exercise, etc. These more subjective aspects of the course proved to be invaluable in drawing the class together and making physical geography more interesting. *Please E-mail the text author if you would like him to send you a copy of any of these exercises, and please feel free to send him examples of your equally valid lecture-class exercises.* The following is just a sampling of several ideas.

• *Student Study Guide.* If you are using the *Student Study Guide* that accompanies *Elemental Geosystems* you could base 30 or 40% of your student's final grade on its completion and the remaining points on objective tests. As an incentive for review, you could have groups of chapters due on the day of each exam. The students tear them out and staple them together for submission. It is easy to make up small strips of paper that list the grades for the study guide chapter work using a simple "Grade: √+ , √ , √− , or 0" system and give them their earned points that go with these marks. The strips are given to the students, but not the study guide pages to them. Using the *Student Study Guide* greatly improves test scores and class discussions! In many instances questions and class interactions are stimulated by the study guide. Here are a few exercise ideas that were replaced by the *Student Study Guide*.

• **Library exercise.** It is amazing how little exposure the students have had to the library both physical and virtual—"that big building in the middle of campus, don't you know?!" Yet facility with the library is key to their academic success. Now with the Internet and learning resource centers on campus, the function of the library is changing. Now to complete the Internet exercise many students are spending time at computers stationed in the library. We recommend that you work with your librarians to refine your exercise. The students, after-the-fact, are constantly referring to the impact the exercise had on them. The items in the four-page handout include work with computer files, computerized magazine and journal summaries, Internet, computerized newspaper summary system, periodicals, pamphlet files, and reference collections. It ends with a request that they get a library card and a list of the library's operating hours—seemingly obvious tasks. On our campus the free library card gives them access to the Learning Resource Center where they can access the Internet and word process on both PC and Macintosh systems.

• **Weather observation exercise.** This is a simple handout for seven days of basic weather observations using the standardized weather station symbols published by the National Weather Service as shown in text Figure 5.33 below an idealized weather map for North America. Included are temperature, pressure, state of the sky, weather type, wind direction and speed, and cloud type. Students need to locate a source for the weather data: the cable Weather Channel, a weather phone number, the many Internet links, or instruments on campus. The students are asked to turn it in whenever the seven days are done, which drives some of them nuts because they want to know exactly when it is due.

• **Climographs and sample station exercise.** Using the basic climograph form as it appears in Chapter 3 and 7 (a blank climograph is in the overhead transparency set for you to use), students graph several stations of their choice spread through a variety of climates, then describe the climate category or Köppen classification and determine the range of air pressures between January and July, as well as the dominant soil order and the appropriate terrestrial biome designation. This requires that they locate their cities in an atlas and transfer that location to the maps in the text. The second part of the exercise asks them to select one of their cities and write a brief geographic analysis of the place and its regional setting; this can include aspects of both physical and human geography.

• **Article analysis exercise.** The purpose of this semester-long exercise is to find and analyze articles that appear in news-papers, magazines, journals, or any other media that relate to the geography course. Students are asked to prepare a couple of pages of analysis for each item. In addition to printed sources, they may also utilize the many sources on the Internet, the Public Broadcasting System (PBS), National Public Radio (NPR), and C-SPAN and such programs as NOVA, the Lehrer News Hour, National Geographic, Nature, "The Morning Report" and "All Things Considered" on NPR. This is assigned the first week of class and is not due until the last week of the term, therefore providing an opportunity to do a couple of minutes of opening warm-up in almost every class using applied topics to hook their interest. In this way, they can see how much of the course content and related issues and controversies are actually occurring out there in the real world. Also, this allows you to point out items in professional journals and related literature, and it encourages students to possibly upgrade their research papers for other classes with geographic content.

The Teacher: Three Aspects

Obviously, those of you who are veteran teachers may have many of these thoughts from your own experience and you probably could have written this section yourself. With this in mind, we offer the following.

The attribute that is the sustaining force behind teaching is an applied common sense, a feel for the appropriateness of a practice. No matter what skills teachers develop, if they do not see, hear, and feel what is happening in the class, then the potential of the moment is defeated and the class will drag painfully through the term. A teacher must develop the ability to imagine the feelings, sensations, and thinking that are going on in the student. This only can be done through being with them in the moment and not being isolated and detached. This is the *empathy force*.

The "eyebrow index" is an extremely good indicator of what is happening in the student. The constant sampling of all the eyes in the room can direct you in your presentation. Frowns and furrows trigger a repeat variation of the material just presented, followed by another sequence of questioning—show, discuss, apply, show, discuss, apply. Admittedly, when you are teaching in a large classroom, such empathetic eye contact is difficult; perhaps if you can move up the aisles it will facilitate seeing their faces. Time does not allow for the resolution of all the negative eyebrows, but movement in the majority signals that new material can be presented. Since students' facial expressions are somewhat unconscious, they are surprised and thrilled when the teacher perceives their need for help.

To intensify the eyebrow indicator, you can refer back to students who previously asked questions or made statements. And if you remember, you can work such referrals in somehow, even if it is later in the semester. This has a great impact on students to see that what they say and do is remembered. This also improves the quality of questions, for they know their questions are taken seriously.

Many characteristics can be listed as applicable to the teacher. This simplified consideration has three idealized categories. Each of these can be thought of as broad paradigms of behavior and attributes, variable in their mix and intensity within each individual.

The Teacher as a Personality. Each of us has our own individual personality and projects a persona or social mask to others, hopefully one that is appropriate for the social role we play in society. When in front of a class, every teacher generates a certain teacher persona that faces out to a roomful of student personas. We believe that the thinner these necessary masks become, the greater the energy that is generated for the teacher-student interaction. For some, merely stepping out from behind the podium or lectern might raise the pulse rate. Sharing an important personal anecdote to heighten a point might quicken your own breathing and heart rate. Experiment with this in your own situation and feel the increase in nervous energy. The challenge is to translate this energy into positive fuel for the presentation.

Allowing your primary person to shine through your mask and learning how to channel the energy thus derived is an important aspect of successful teaching. The student can see a real personality in all its glory, perfections and flaws and the works. There are few thrills that compare with that rush of goose-bumps one gets when explaining some majestic point about Earth's physical geography to receptive faces—especially when it's for the umpteenth time after umpteen years of teaching! It's

all right to feel emotion and instill emotion over these dramatic environmental subjects. After all, this is a dramatic and beautiful planet and students do respond to passion about it!

The Teacher as Intellect. Intelligence is an especially important trait of a teacher. Critical to the craft of teaching is this power to know and understand, for the teacher must comprehend information specific to the study at hand. You may have experienced insights and new learning yourself while right in the middle of lecturing on an "old" topic. These flashes are always interesting and certainly keep one humble, because you never know when the lightning is going to strike! To have up-to-date material, constantly in revision, is crucial to teaching, and the freshness that is derived is one of the rewards of continuing learning. The effective teaching of information is different from researching and publishing that information.

Teachers of physical and cultural geography must be eclectic, that is, able to draw from many disciplines as well as from their own field. The ability to synthesize requires patterned seeing and relational thinking, if notes for the classroom discussion are to be dynamic and full of energy.

The Teacher as Actor. Teaching is a performing art and therefore involves a performance. Voice is very important—enunciation, modulation, and projection. Direct your voice outwards, not into the lectern. And remember, what seem like exaggerated gestures and voice patterns will seem like normal speech in the back of the room, whereas a normal voice from the lecturer will come across as flat and sterile. The "actor" role is what handles the repetition of content in performance to each new "audience."

You might want to practice: using a recorder, tape your voice for a minute or two, reading, lecturing, and speaking; then, with that feedback, adjust accordingly. Perhaps you could have a lecture-discussion videotaped for review. (Note the nervous energy when you go to push the record button—tap into that!)

Importantly, because the goal is communication and learning, the performance should not be for its own sake but be a means to the educational objectives in operation: your preparation and effort is a catalytic vehicle. With such an exciting discipline as physical geography, we must watch out for excess however; "Gee, that teacher is entertaining but we don't learn much," is probably not the response we're looking for.

There is constant pressure in the classroom for substance in content and presentation, acting almost like positive feedback in a geographic system. Classroom teaching seems capable of absorbing any increase in energy input—seemingly without limit! The possibilities are infinite, and with Earth as the subject, the challenges are exhilarating.

For Ready Reference

The table of contents and complete focus study and news report list are printed here for your convenience.

The "Geography I. D." Assignment for Students:

GEOGRAPHY I.D.

To begin: complete your personal *Geography I. D.* Use the maps in your text, an atlas, college catalog, and additional library materials if needed.

Your instructor will help you find additional source materials for data pertaining to the campus.

On each map find the information requested noting the January and July temperatures indicated by the isotherms (the small scale of these maps will permit only a general determination), January and July pressures indicated by isobars, annual precipitation indicated by isohyets, climatic region, landform class, soil order, and ideal terrestrial biome.

Record the information from these maps in the spaces provided. The completed page will give you a relevant geographic profile of your immediate environment. As you progress through your physical geography class the full meaning of these descriptions will unfold. This page might be one you will want to keep for future reference.

[* Derived from *Geosystems Student Study Guide*, sixth edition, by Robert Christopherson and Charles E. Thomsen, and *Applied Physical Geography: Geosystems in the Laboratory*, sixth edition, by Robert W. Christopherson and Charles E. Thomsen, © 2006 Prentice Hall, Inc., Pearson Education]

Geography I.D.

Name: _____ Class Section: _____

Home Town: _____ Latitude: _____ Longitude: _____

College/University: _____

City/Town: _____ County (Parish): _____

Standard Time zone (for College Location): _____

Latitude: _____ Longitude: _____

Elevation (include location of measurement on campus): _____

Place (tangible and intangible aspects that make this place unique): _____

Region (aspects of unity shared with the area; cultural, historical, economic, environmental): _____

_____

Population: Metropolitan Statistical Area (CMSA, PMSA, if applicable): _____

Environmental Data: (Information sources used: _____)

    January Avg. Temperature: _____ July Avg. Temperature: _____

    January Avg. Pressure (mb): _____ July Avg. Pressure: _____

    Average Annual Precipitation (cm/in.): _____

    Avg. Ann. Potential Evapotranspiration (if available; cm/in.): _____

Climate Region (Köppen symbol and name description): _____

Main climatic influences (air mass source regions, air pressure, offshore ocean temperatures, etc.) _____

_____

_____

Topographic Region or Structural Region (inc., rocks type, loess units, etc.): _____

_____

Dominant Regional Soil Order _____

Biome (terrestrial ecosystems description; ideal and present land use): _____

_____

_____

Foundations of Geography

Our study of physical geography begins with the foundations of this important discipline. The "Foundations of Geography" chapter contains the basic tools for the student to use in studying the content of physical geography. After completion of this chapter you should feel free to follow the integrated sequence of the text or treat the four parts of the text in any order that fits your teaching approach.

Most students have different notions as to what geography is and what geographers do. Some think they are going to memorize the capitals of the states, while others think that they are embarking on a search for lands and peoples or that they will be learning things for success with a popular *Jeopardy!* category. This uncertainty as to the nature of geography provides an excellent opportunity to review the state of affairs relative to geographic awareness. For information about geography as a discipline and career possibilities contact the Association of American Geographers and ask for their pamphlets, "Careers in Geography" by Richard G. Boehm (1996), "Geography: Today's Career for Tomorrow," "Geography as A Discipline" by Richard E. Huke (1995), and "Why Geography?" also revised in 1995. Order by phone (1.202-234-1450, or E-mail at gaia@aag.org, or see their home page at **http://www.aag.org**.

In my own physical and cultural geography classes I find that many students are unable to name all the states and provinces or identify major countries on outline maps. When we add the complexity of the spatial aspects of Earth's physical systems, (i.e., atmospheric energy budgets, temperatures, wind patterns, weather systems, plate tectonics, earthquake and volcano locations and causes, global ecosystem diversity, and terrestrial biomes), the confusion even among the best informed is great and therein lies our challenge!

These are certainly dramatic times, with an expanding global concern for the environment moving to the forefront of the political and scientific agenda. In 1992, the United Nations Conference on Environment and Development (UNCED) held an international gathering of scientists, political leaders, and citizens in

Rio de Janeiro. A key portion of the conference dealt with issues central to physical geography and our human-environment theme: preservation of biological diversity, climate change and global warming, stratospheric ozone depletion, trans-boundary air pollution, deforestation, soil erosion and loss, desertification, permanent drought, freshwater resources, and resources from the ocean and coastlines. In addition to these environmental topics, issues relating to human societies also were addressed. Full international attention improves as negative feedback from Earth's physical environment continues to draw the world together—but this only can take root with a strong foundation in physical geography.

Outline Headings and Key Terms

The first-, second-, and third-order headings that divide Chapter 1 serve as an outline for your notes and studies. The key terms and concepts that appear boldface in the text are listed here under their appropriate heading in bold italics. All these highlighted terms appear in the text glossary. Note the check-off box (❑) so you can mark your progress as you master each concept. Your students have this same outline in their *Student Study Guide*.

The outline headings for Chapter 1:

   ❑ *Earth systems science*
The Science of Geography
   ❑ *geography*
   ❑ *spatial*
   ❑ *location*
   ❑ *place*
   ❑ *movement*
   ❑ *region*
   ❑ *human–Earth relationships*
Geographic Analysis
   ❑ *spatial analysis*
   ❑ *process*

The URLs related to this chapter of *Elemental Geosystems* can be found at
http://www.prenticehall.com/christopherson

Key Learning Concepts

The following key learning concepts help guide your reading and comprehension efforts. The operative words are in *italics*. Use these carefully to guide your reading of the chapter and note that STEP 1 asks you to work with these concepts. The same key learning concepts open the chapter on its title page.

After reading the chapter and using this study guide, you should be able to do the following:

• *Define* geography and physical geography in particular.
• *Describe* systems analysis, open and closed systems, feedback information, and system operations, and *relate* these concepts to Earth systems.
• *Explain* Earth's reference grid: latitude and longitude, plus latitudinal geographic zones and time.

• *Define* cartography and mapping basics: map scale and map projections.
• *Describe* remote sensing and *explain* geographic information system (GIS) as a tool used in geographic analysis.

Annotated Chapter Review Questions

• *Define* **geography, and physical geography in particular.**

1. What is unique about the science of geography? On the basis of information in this chapter, define physical geography and review the geographic approach.

Geography is the science that studies the interdependence of geographic areas, places, and locations; natural systems; processes; and societal and cultural activities over Earth's surface. Physical geography involves the spatial analysis of Earth's physical environment. Various words denote the geographic context of spatial analysis: space, territory, zone, pattern, distribution, place, location, region, sphere, province, and distance. Spatial patterns of Earth's weather, climate, winds and ocean currents, topography, and terrestrial biomes are examples of geographic topics.

2. In general terms, how might a physical geographer analyze water pollution in the Great Lakes?

There are many ways to answer here. First get map coverage of the Great Lakes Region (see the Focus Study on the Great Lakes in Chapter 19). Describe the lake elevations, flows, volumes, and annual mixing patterns as temperatures change seasonally. Locate population centers and point sources of pollution and using population concentrations estimate nonpoint sources of pollution. Map published data of water chemical analyses. Using a GIS model develop a composite overlay of all the above elements.

3. Assess your geographic literacy by examining atlases and maps. What types of maps have you used—political? physical? topographic? Do you know what projections they employed? Do you know the names and locations of the four oceans, seven continents, and individual countries? Can you identify the new countries that have emerged since 1990?

An informal quiz might be appropriate here, as in, Ouagadougou is the capital of Burkina, formerly Burkina Faso. Where are Kuwait and Iraq and how many live there? What is Burma's proper name today? What new nations have formed since 1990? What is the current spelling and name for Bombay, India? Where are warm and wet, hot and dry, cool and moist, cold and dry regions? Where are you right now in terms of latitude, longitude, elevation, etc.?

4. Suggest a representative example for each of the five geographic themes and use that theme in a sentence.

Suggestion: use Figure 1.1, text and photographs, as cues for discussion of this question. The Association of American Geographers (AAG) and the National Council for Geographic Education (NCGE), in an attempt to categorize the discipline, set forth five key themes for modern geographic education: location, place, human-Earth relationships, movement, and region.

5. Have you made decisions today that involve geographic concepts discussed within the five themes presented? Explain briefly.

This may be a good assignment for the first day of class. Most students can easily complete a history, or chronological list, of a day's activities. Have students do a geographic journal of their day so far. Listing locations they have been, how they got there and have them draw a map of their journey from home to school to work. This will also demonstrate how cartographers attempt to select information to include in their maps and how geographers attempt to select the most significant variables in a systems analysis.

• *Describe* systems analysis and open and closed systems, and *relate* these concepts to Earth systems.

6. Define systems theory as an organizational strategy. What are open systems, closed systems, and negative feedback? When is a system in a steady-state equilibrium condition? What type of system (open or closed) is a human body? A lake? A wheat plant?

Simply stated, a system is any ordered, interrelated set of objects, things, components, or parts, and their attributes, as distinct from their surrounding environment. A natural system is generally not self-contained: inputs of energy and matter flow into the system, whereas outputs flow from the system. Such a system is referred to as an open system: the human body, a lake, or a wheat plant. A system that is shut off from the surrounding environment so that it is self-contained is known as a closed system.

As a system operates, information is generated in the system output that can influence continuing system operations. These return pathways of information are called feedback loops. Feedback can cause changes that in turn guide further system operations. If the information amplifies or encourages responses in the system, it is called positive feedback. On the other hand, negative feedback tends to slow or discourage response in the system, forming the basis for self-regulation in natural systems and regulating the system within a range of tolerable performance. When the rates of inputs and outputs in the system are equal and the amounts of energy and matter in storage within the system are constant, or, more realistically, as they fluctuate around a stable average, the system is in a steady-state equilibrium.

7. Describe Earth as a system in terms of energy and of matter.

Most systems are dynamic because of the tremendous infusion of radiant energy from reactions deep within the Sun. This energy penetrates Earth's atmosphere and cascades through the terrestrial systems, transforming along the way into various forms of energy. Earth is an open system in terms of energy. In terms of physical matter—air, water, and material resources—Earth is nearly a closed system.

8. What are the three abiotic (nonliving) spheres that make up Earth's environment? Relate these to the biotic sphere (living): the biosphere.

Earth's surface is the place where four immense open systems interface, or interact. Three nonliving, or abiotic, systems overlap to form the realm of the living, or biotic, system. The abiotic spheres include the atmosphere, hydrosphere, and lithosphere. The fourth, the biotic sphere, is called the biosphere and exists in an interactive position between and within the abiotic spheres.

• *Explain* Earth's reference grid: latitude and longitude, and latitudinal geographic zones and time.
9. Draw and describe Earth's shape and size.

Sketch after Figure 1.8 an oblate spheroid or geoid.

10. How did Eratosthenes use Sun angles to figure out that the 5000 stadia distance between Alexandria and Syene was 1/50 Earth circumference? Once he knew this fraction of Earth's circumference, how did he calculate the distance of Earth's circumference?

See Figure 1.9 and label those parts that demonstrate the methods Eratosthenes used to do his calculation.

11. What are the latitude and longitude coordinates (in degrees, minutes, and seconds) of your present location? Where can you find this information?

Have students find a local or library source: benchmarks, atlases, gazetteers, maps, many software

programs, such as Global Explorer, and locations on the World Wide Web.

12. Define latitude and parallel and define longitude and meridian using a simple sketch with labels.

On a map or globe, lines denoting angles of latitude run east and west, parallel to Earth's equator. Latitude is an angular distance north or south of the equator measured from a point at the center of Earth. A line connecting all points along the same latitudinal angle is called a parallel. (Figures 1.10 and 1.11.)

On a map or globe, lines designating angles of longitude run north and south at right angles (90°) to the equator and all parallels. Longitude is an angular distance east or west of a surface location measured from a point at the center of Earth. A line connecting all points along the same longitude is called a meridian. (Figure 1.12.)

13. Define a great circle, great circle routes, and a small circle. In terms of these concepts, describe the equator, other parallels, and meridians.

A great circle is any circle of Earth's circumference whose center coincides with the center of Earth. Every meridian is one-half of a great circle that crosses each parallel at right angles and passes through the poles. An infinite number of great circles can be drawn on Earth, but only one parallel is a great circle—the equatorial parallel. All the rest of the parallels diminish in length toward the poles, and, along with other circles that do not share Earth's center, constitute small circles.

14. Identify the various latitudinal geographic zones that roughly subdivide Earth's surface. In which zone do you live?

Figure 1.11 portrays these latitudinal geographic zones, their locations, and their names: equatorial, tropical, subtropical, midlatitude, subarctic or subantarctic, and Arctic or Antarctic.

15. What does timekeeping have to do with longitude? Explain their relationship. How is Coordinated Universal Time (UTC) determined on Earth?

Earth revolves 360° every 24 hours, or 15° per hour, and a time zone of one hour is established for each 15° of longitude. Thus, a world standard was established, and time was set with the prime meridian at Greenwich, England. Each time zone theoretically covers 7.5° on either side of a controlling meridian and represents one hour. Greenwich Mean Time (GMT) is called Coordinated Universal Time (UTC); and although the prime meridian is still at Greenwich, UTC is based on average time calculations kept in Paris and broadcast worldwide. UTC is measured today by the very regular vibrations of cesium atoms in 6 primary standard clocks—the *NIST-7* being the newest placed in operation during 1994 by the United States. A newer clock for the US went into operation in 2000, designated *NIST-F1*.

16. What and where is the prime meridian? How was the location originally selected? Describe the meridian that is opposite the prime meridian on the Earth's surface.

In 1884, the International Meridian Conference was held in Washington, D.C. After lengthy debate, most participating nations chose the Royal Observatory at Greenwich as the place for the prime meridian of 0° longitude. An important corollary of the prime meridian is the location of the 180° meridian on the opposite side of the planet. Termed the International Date Line, this meridian marks the place where each day officially begins (12:01 A.M.) and sweeps westward across Earth. This westward movement of time is created by the planet's turning eastward on its axis.

17. What is GPS and how does it assist you in finding location and elevation on Earth? Give a couple of examples where it was utilized to correct heights for some famous mountains.

The Global Positioning System (GPS) comprises 24 orbiting satellites, in 6 orbital planes, that transmit navigational signals for Earth-bound use (backup GPS satellites are in orbital storage as replacements). A small receiver, some about the size of a pocket radio, receives signals from four or more satellites at the same time, calculates latitude and longitude within 10-m accuracy (33 ft) and elevation within 15 m (49 ft) and displays the results. With the shutdown in 2000 of the Pentagon Selective Availability, commercial resolution is the same as for military applications and its Precise Positioning Service (PPS).

Scientists used GPS to accurately determine the height of Mount Everest in the Himalayan Mountains—now 8850 m compared to the former 8848 m (29,035 ft, 29,028 ft). In contrast GPS measurements of Mount Kilimanjaro lowered its summit from 5895 m to a lower 5892 m (19,340 ft, 19,330 ft).

• *Define* cartography and mapping basics: map scale and map projections.

18. Define cartography. Explain why it is an integrative discipline.

The part of geography that embodies map making is called cartography. The making of maps and charts is a specialized science as well as an art, blending aspects of geography, engineering, mathematics, graphics, computer sciences, and artistic specialties. It is similar in ways to architecture, in which aesthetics and utility are combined to produce an end product.

19. What is map scale? In what three ways is it expressed on a map?

The expression of the ratio of a map to the real world is called scale; it relates a unit on the map to a similar unit on the ground. A 1:1 scale would mean that a centimeter on the map represents a centimeter on the ground—certainly an impractical map scale, for the map would be as large as the area being mapped. A more appropriate scale for a local map is 1:24,000. Map scales may be presented in several ways: written, graphic, or as a representative fraction.

20. State whether each of the following ratios is a large scale, medium scale, or small scale: 1:3,168,000, 1:24,000, 1:250,000.

Small, large, and intermediate scales (see Table 1.2, p. 21).

21. Describe the differences between the characteristics of a globe and those that result when a flat map is prepared.

Since a globe is the only true representation of distance, direction, area, shape, and proximity, the preparation of a flat version means that decisions must be made as to the type and amount of acceptable distortion. On a globe, parallels are always parallel to each other, evenly spaced along meridians, and decrease in length toward the poles. On a globe, meridians intersect at both poles and are evenly spaced along any individual parallel. The distance between meridians decreases toward poles, with the spacing between meridians at the 60th parallel being equal to one-half the equatorial spacing. In addition, parallels and meridians on a globe always cross each other at right angles. All these qualities cannot be reproduced on a flat surface. Flat maps always possess some degree of distortion.

22. What type of map projection is used in Figure 1.11? Figure 1.15?

Figure 1.11: a Robinson projection; Figure 1.15: a Mercator projection (see Appendix A for discussion of maps used in this text).

• *Describe* remote sensing and *explain* geographic information system (GIS) as a tool used in geographic analysis.

23. What is remote sensing? What are you viewing when you observe a weather image on TV or in the newspaper? Explain.

Our eyes and cameras are familiar means of obtaining **remote-sensing** information about a distant subject without having physical contact. Remote sensors on satellites and other craft sense a broader range of wavelengths than can our eyes. They can be designed to "see" wavelengths shorter than visible light (ultraviolet) and wavelengths longer than visible light (infrared and microwave radar).

Active systems direct a beam of energy at a surface and analyze the energy that is reflected back. An example is *radar* (*ra*dio *d*etection *a*nd *r*anging). Passive remote-sensing systems record energy radiated from a surface, particularly visible light and infrared.

A new generation of the Geostationary Operational Environmental Satellite, known as *GOES*, became operational in late 1994 providing frequent infrared and visible images—the ones you see on television weather reports. *GOES-10*, new in 1998, is positioned above 135° W longitude to monitor the West Coast and the eastern Pacific Ocean. *GOES-8* is positioned above 75° W longitude to monitor central and eastern North America and the western Atlantic. The television and newspaper weather images we see are from these platforms.

24. Describe *Terra*, *Landsat*, *GOES*, and *SPOT*, and explain them using several examples.

Key to NASA's Earth Observing System (EOS) is satellite *Terra*, which began beaming back data and images in 2000 (see **http://terra.nasa.gov/**), this followed by another satellite in the series called *Aqua*. Five instrument packages observe Earth systems in detail, exploring the atmosphere, landscapes, oceans, environmental change, and climate, among other abilities. For example, the Clouds and the Earth's Radiant Energy System (CERES) instruments aboard *Terra* monitor the Earth's energy balance, giving new insights into climate change (see Chapter 4). These monitors offer the most accurate global radiation and energy measurements ever available. Another instrument set, the Moderate-resolution Imaging Spectroradiometer (MODIS) sees all of Earth's surface every 1–2 days in thirty-six spectral bands, thereby expanding on AVHRR capabilities. There are 29 *Terra* images, out of 105 remotely sensed images in this edition of *Geosystems*.

Passive remote sensors on five *Landsat* satellites, launched by the United States, provide a variety of data as shown in images of river deltas in Chapter 14, the Malaspina glacier in Alaska in Chapter 17, and the Great Lakes in Chapter 19. Three *Landsats* remain operational (4, 5, and the newest 7) although *Landsat 4* no longer gathers images and is used for orbital tests. *Landsat 5* remains threatened with shut off due to budget cuts.

One of two commercial systems includes the three French satellites (numbered 1, 2, and 4) called *SPOT* (Systeme Probatoire d'Observation de la Terre; see **http://spot4.cnes.fr/waiting.htm**), that resolve objects on Earth down to 10 to 20 m (33 to 66 ft), depending on which sensors are used.

The *GOES* description is in the previous annotated question.

25. If you were in charge of planning for development of a large tract of land, how would GIS methodologies assist you? How might planning and zoning be affected if a portion of the tract in the GIS is a floodplain or prime agricultural land?

GIS enables you to gather, manipulate, and analyze vast amounts of geographic information from remote sensing or satellite technologies. The coordinates of a specific location may be digitized, and then geographers may add layers of remote sensing information to these coordinates. This allows geographers to overlay analytical information from more than one data plane and analyze complex relationships rapidly. This enables geographers to manipulate variables within their study very rapidly.

The usefulness of GIS in analyzing a floodplain would be that the user could examine the frequency of flooding, and the areal or monetary extent of damage in the last ten or more years. For agricultural land, the user could examine levels of salinity, soil fertility, crop production statistics for previous crops, and locate areas within their land that may need extra care or different methods of farming. GIS is applicable to urban and transport planning because it is able to consider and model areas of hazard, agricultural land values, and routes of intensive transport use.

Overhead Transparencies

	Chapter 1	Foundations of Geography
1.	CO 1	Cape Hatteras Light, plus inset photo
2.	1.1	Five themes of geographic science
3.	F.S. 1.1.1	Scientific method flow chart
4.	1.2	The content of geography
5.	1.3 & 1.4	An open system, top; A leaf is a natural open system, bottom
6.	1.5	Global impact of Mount Pinatubo eruption
7.	1.6	The systems in *Elemental Geosystems*
8.	1.7	Earth's four spheres
9.	1.8 and 1.9	Earth's dimensions (top); Eratosthenes' calculation (bottom)
10.	1.10 and 1.12	Parallels of latitude; Meridians of longitude
11.	1.11	Latitudinal geographic zones
12.	1.13	Great circles and small circles
13.	1.15	Modern international standard time zones
14.	1.16	International Date Line location
15.	1.18	From globe to a flat map
16.	1.19 a,b,c,d	Classes of map projections
17.	1.20 a, b	Gnomonic, great circle routes; Mercator, true compass direction
18.	1.21	Remote-sensing technologies
19.	1.26 a,b,c	A geographic information system model

PART ONE:
The Energy-Atmosphere System

Overview

Elemental Geosystems begins with the Sun and solar system to launch the first of four parts. Our planet and our lives are powered by radiant energy from the star closest to Earth—the Sun. Each of us depends on many systems that are set into motion by energy from the Sun. These systems are the subjects of Part One.

Part One exemplifies the systems organization of the text: it begins with the origin of the solar system and the Sun. Solar energy passes across space to Earth's atmosphere, varying seasonally in its effects on the atmosphere. Insolation then passes through the atmosphere to Earth's surface. From Earth's surface, atmospheric and surface energy balances generate patterns of world temperature and general and local atmospheric circulations. Each part contains related chapters with content arranged according to the flow of individual systems or in a manner consistent with time and the flow of events.

The Part-opening photo shows sunrise in the Scoresby Sund fjord system, East Greenland in September 2003, at approximately 71° N latitude. The extended length of dawn and twilight in the high latitudes produces wonderful colors and long shadows from transient icebergs. The icebergs probably calved and set sail from glaciers in northern Greenland, were ocean going driven by winds, and floated into this fjord system—all involving physical principles in PART One, The Energy-Atmosphere System.

Solar Energy, Seasons, and the Atmosphere

The ultimate spatial inquiry is to discern the location of Earth in the Universe. To properly set the stage for a course in the physical geography of Earth, slides, videography, and posters can be used to establish the location and place of our planetary home. Our immediate home is North America, a major continent on planet Earth, the third planet from a typical yellow star in a solar system. That star, our Sun, is only one of billions in the Milky Way Galaxy, which is one of millions of galaxies in the Universe.

This chapter examines the nature of the flow of energy and material from the Sun to the outer reaches of Earth's atmosphere. Earth–Sun orbital relations during the year produce the pulse of seasons and the annual changes in daylength, Sun altitude and declination. The passage of solar energy through Earth's atmospheric layers is profiled, as is the filtering of harmful radiation.

The chapter examines the deterioration of valuable stratospheric ozone as a result of human-produced gases. Anthropogenic gases—air pollution—affects life, economies, and nature. Policies such as the Clean Air Act dramatically turned the air pollution problem around. CAA benefits outweigh costs by at least 42 to 1.

Outline Headings and Key Terms

The first-, second-, and third-order headings that divide Chapter 2 serve as an outline for your notes and studies. The key terms and concepts that appear boldface in the text are listed here under their appropriate heading in bold italics. All these highlighted terms appear in the text glossary. Note the check-off box (❑) so you can mark your progress as you master each concept. Your students have this same outline in their *Student Study Guide*. The ☉ icon indicates that there is an accompanying animation or other resource on the CD.

The outline headings for Chapter 2:

The Solar System, Sun, and Earth
☉ **Nebular Hypothesis**
- [ ] *Milky Way Galaxy*
- [ ] *planetesimal hypothesis*

Dimensions, Distances, and Earth's Orbit
- [ ] *speed of light*
- [ ] *perihelion*
- [ ] *aphelion*

Solar Energy: From Sun to Earth
☉ **Electromagnetic Spectrum and Plants**
- [ ] *fusion*

Solar Wind
- [ ] *solar wind*
- [ ] *sunspots*
- [ ] *magnetosphere*
- [ ] *auroras*

Electromagnetic Spectrum of Radiant Energy
- [ ] *electromagnetic spectrum*
- [ ] *wavelength*

Incoming Energy at the Top of the Atmosphere
- [ ] *thermopause*

Solar Constant
- [ ] *insolation*
- [ ] *solar constant*

Uneven Distribution of Insolation
- [ ] *subsolar point*

Global Net Radiation

The Seasons
☉ **Earth–Sun Relations, Seasons**
Seasonality
- [ ] *altitude*
- [ ] *declination*
- [ ] *daylength*

Reasons for Seasons
Revolution
- [ ] *revolution*

Rotation
- [ ] *rotation*
- [ ] *circle of illumination*

Tilt of Earth's Axis
- [ ] *plane of the ecliptic*

Axial Parallelism
- [ ] *axial parallelism*

Annual March of the Seasons
- [ ] *winter solstice*
- [ ] *December solstice*
- [ ] *tropic of Capricorn*
- [ ] *vernal equinox*
- [ ] *March equinox*
- [ ] *summer solstice*
- [ ] *June solstice*

- [ ] *tropic of Cancer*
- [ ] *autumnal equinox*
- [ ] *September equinox*

Seasonal Observations

Atmospheric Composition, Temperature and Function
☉ **Ozone Breakdown, Ozone Hole**
- [ ] *exosphere*

Atmospheric Profile
- [ ] *air pressure*

Atmospheric Composition Criterion
Heterosphere
- [ ] *heterosphere*

Homosphere
- [ ] *homosphere*

Atmospheric Temperature Criterion
Thermosphere
- [ ] *thermosphere*
- [ ] *kinetic energy*
- [ ] *sensible heat*

Mesosphere
- [ ] *mesosphere*

Stratosphere
- [ ] *stratosphere*

Troposphere
- [ ] *troposphere*
- [ ] *normal lapse rate*
- [ ] *environmental lapse rate*

Atmospheric Function Criterion
Ionosphere
- [ ] *ionosphere*

Ozonosphere
- [ ] *ozonosphere*
- [ ] *ozone layer*
- [ ] *chlorofluorocarbons* (CFCs)

Variable Atmospheric Components
Natural Sources
Natural Factors that Affect Air Pollution
Wind
Local and Regional Landscapes
Temperature Inversion
- [ ] *temperature inversion*

Anthropogenic Pollution
Photochemical Smog Pollution
- [ ] *photochemical smog*
- [ ] *peroxyacetyl nitrates (PAN)*

Industrial Smog and Sulfur Oxides
- [ ] *industrial smog*
- [ ] *sulfur dioxide*
- [ ] *sulfate aerosols*

Particulates
- [ ] *particulate matter (PM)*
- [ ] *anthropogenic atmosphere*

Benefits of the Clean Air Act
Summary and Review
News Reports and Focus Study

News Report 2.1: Falling Through the Atmosphere—
  the Highest Sky Dive
Focus Study 2.1: Stratospheric Ozone Losses: A
  Worldwide Health Hazard
✧ Ozone Breakdown, Ozone Hole
✧ Satellite loop: Southern Hemisphere 2002–2003
Focus Study 2.2: Acid Deposition: A Continuing Blight
  on the Landscape

The URLs related to this chapter of *Elemental
Geosystems* can be found at
http://www.prenticehall.com/christopherson

Key Learning Concepts

After reading the chapter and using the study
guide, the student should be able to:

• *Distinguish* among galaxies, stars, and planets and
locate Earth.
• *Describe* the Sun's operation and *explain* the
characteristics of the solar wind and the
electromagnetic spectrum of radiant energy.
• *Define* solar altitude, solar declination, and daylength
and *describe* the annual variability of each—Earth's
seasonality.
• *Construct* a general model of the atmosphere based on
composition, temperature, and function and *diagram*
this model in a simple sketch.
• *Describe* conditions within the stratosphere;
specifically, *review* the function and status of the
ozonosphere (ozone layer).
• *Distinguish* between natural and anthropogenic
variable gases and materials in the lower atmosphere
and *describe* the sources and effects of air pollution and
acid deposition.

Annotated Chapter Review Questions

• *Distinguish* **among galaxies, stars, and planets, and**
locate **Earth.**

**1. Describe the Sun's status among stars in the
Milky Way Galaxy. Describe the Sun's location, size,
and relationship to its planets.**

Our Sun is both unique to us and
commonplace in our galaxy. It is only average in
temperature, size, and color when compared with other
stars, yet it is the ultimate energy source for almost all
life processes in our biosphere. Planets do not produce
their own energy. Our Sun is located on a remote,
trailing edge of the Milky Way Galaxy, a flattened,
disk-shaped mass estimated to contain up to 400 billion
stars.

An excellent poster for this discussion comes
as a supplement with the June 1983 *National
Geographic* magazine (Vol. 163, No. 6, p. 704A),
"Journey into the Universe Through Time and Space."
(Back issues are available from NGS.) The same
illustration appears modified in the editions of the *Atlas
of the World* by National Geographic, now in its
seventh edition (p. 24, "The Universe, Nature's
Grandest Design").

**2. If you have seen the Milky Way at night, briefly
describe it. Use specifics from the text in your
description.**

From our Earth-bound perspective in the
Milky Way, the galaxy appears to stretch across the
night sky like a narrow band of hazy light. On a clear
night the naked eye can see only a few thousand of the
nearly 400 billion stars.

**3. Briefly describe Earth's origin as part of the Solar
System.**

According to prevailing theory, our solar
system condensed from a large, slowly rotating,
collapsing cloud of dust and gas called a nebula. As the
nebular cloud organized and flattened into a disk shape,
the early proto-Sun grew in mass at the center, drawing
more matter to it. Small accretion (growing) eddies—
the protoplanets—swirled at varying distances from the
center of the solar nebula. The early protoplanets, or
planetesimals, were located at approximately the same
distances from the Sun that the planets are today. The
beginnings of the Sun and the solar system are
estimated to have occurred more than 4.6 billion years
ago. These processes are now observed as occurring
elsewhere in the galaxy. Astronomers so far have
observed almost two dozen stars with planets orbiting
about them.

**4. Compare the nine planets of the Solar System and
their distances from the Sun.**

Examine Figure 2.1, p. 37. A comparison of
Earth with the other planets is interesting. The average
distance from the Sun to Earth is referred to as an
astronomical unit, or A.U., and is used as a basic
measurement unit in the Solar System. The average
distance from the Sun in terms of astronomical units are
as follows: Mercury = 0.39, Venus = 0.72, Earth = 1.0,
Mars = 1.52, Jupiter = 5.20, Saturn = 9.54, Uranus =
19.18, Neptune = 30.70, and Pluto 39.46. Note: Pluto's
motion is so skewed from the plane of the ecliptic and
eccentric in orbit that Neptune actually is further from
the Sun than Pluto from 1979 until 1999. In 1999, Pluto
crossed the orbit of Neptune and became the most
distant planet again.

5. How far is Earth from the Sun in terms of light speed? In terms of kilometers and miles? Relate this distance to the shape of Earth's orbit during the year.

Earth's orbit around the Sun is presently elliptical—a closed, oval-shaped path (Figure 2.1d). Earth's average distance from the Sun is approximately 150 million km (93 million mi). Earth is at perihelion, its closest position to the Sun, on January 3 at 147,255,000 km (91,500,000 mi) and at aphelion, its farthest distance from the Sun, on July 4 at 152,083,000 km (94,500,000 mi). This seasonal difference in distance from the Sun results in a variation of 3.4 percent in the solar output that is intercepted by Earth.

6. Briefly describe the relationship among these concepts: Universe, Milky Way Galaxy, Solar System, Sun, and Planet Earth.

The Universe is infinite in size and dimension from our perspective. Tens of billions of galaxies are known to populate the Universe we can observe, each composed of hundreds of billions of stars. Our Sun is a typical yellow-dwarf thermonuclear (fusion) star, somewhat less than a million miles in diameter. Planets orbit about stars such as our Solar System. Our Sun and the orbiting planets are revolving around the Milky Way Galaxy in a vast clockwise spiral. Similar multiple-planet solar systems are now being studied through the Hubble and other telescopes.

7. What is the role of gravity, and the gravitational force, in the Solar System and in planetary formation?

According to prevailing theory, our Solar System condensed from a large, slowly rotating, collapsing cloud of dust and gas called a *nebula*. Gravity, the mutual attracting force exerted by the mass of an object upon all other objects, was the key organizing force in this condensing solar nebula. The beginnings of the formation of the Sun and its Solar System are estimated to have occurred more than 4.6 billion years ago.

The concept that suns condense from nebular clouds and planetesimals form in orbits around their central masses is called the **planetesimal hypothesis**, or *dust-cloud hypothesis*. The planets appear to accrete (growth by accumulation) from dust, gases, and icy comets that are drawn by gravity into collision and coalescence. Astronomers are observing this formation process under way in other parts of the galaxy. By 2005, astronomers had identified more than 230 planets orbiting other stars.

• *Describe* the Sun's operation, and *explain* the characteristics of the solar wind and the electromagnetic spectrum of radiant energy.

8. How does the Sun produce such tremendous quantities of energy?

The solar mass produces tremendous pressure and high temperatures deep in its dense interior region. Under these conditions, pairs of hydrogen nuclei, the lightest of all the natural elements, are forced to fuse together. This process of forcibly joining positively charged nuclei is called fusion. In the fusion reaction hydrogen nuclei form helium, the second lightest element in nature, and liberate enormous quantities of energy in the form of free protons, neutrons, and electrons. During each second of operation, the Sun consumes 657 million tons of hydrogen, converting it into 652.5 million tons of helium. The difference of 4.5 million tons is the quantity that is converted directly to energy—resulting in literally disappearing solar mass.

Notes on the Essence of Matter and Fusion. A few basics might be of assistance to support possible questions. An **atom** is the smallest particle of matter that still has distinct properties and can enter into chemical combination. The central core of an atom is called the **nucleus** and is composed of tightly packed particles called **protons** (positive charge) and **neutrons** (neutral charge). Many other smaller particles reside in the nucleus as well. **Electrons** are small, almost weightless, negative electrical charges (probabilistic energy potentials in quantum physics) that exist about the nucleus at varying distances (energy levels). The nucleus in the center of the atom is surrounded by vast amounts of empty space. A neutral atom has the same number of electrons and protons, thereby creating a balanced neutral charge. An **ion** is an ionized atom that has an electric charge with a gained or lost electron. Hydrogen in its most common form (isotope) has one electron and one proton. Oxygen possesses eight protons with eight orbiting electrons, and so on up the periodic table of elements. Two or more atoms bound together form a **molecule**; molecules together form **compounds**.

Hydrogen has one proton and one electron in its principal form; add a neutron to the nucleus and it becomes a **deuterium** hydrogen atom. Helium has two protons and two electrons. Under the conditions of temperature and pressure in the Sun's core, pairs of hydrogen nuclei, the lightest of all the natural elements, are forced to fuse together in various isotopic combinations. This process of forcibly joining positively charged nuclei is called **fusion**. An unstable form of helium, Helium-3, is formed and a neutron is liberated as energy. Helium-3 fuses further with another deuterium to form Helium-4, a stable form of helium, and in this case a proton is liberated as energy. The Sun has converted about 8 percent of its mass to helium overall. The text gives a simple equation of the fusion reaction with hydrogen nuclei forming helium, the second lightest element in nature, and liberating enormous quantities of energy in the form of free protons, neutrons, and electrons that are propelled out from the core of the Sun. The liberated energy takes approximately 10 million years to migrate through the Sun's gaseous densities to its surface!

Every second a tremendous conversion of hydrogen to helium and liberated energy takes place. During each second of operation the Sun consumes 657 million tons of hydrogen, converting it into 652.5 million tons of helium. The difference of 4.5 million tons is the quantity that is converted directly to energy– literally disappearing solar mass just as Einstein postulated in his $E = m \cdot c2$ equation, where the nuclear energy [E] in matter can be calculated for a given mass [m] by multiplying mass times the speed of light [c] squared. Energy and matter are interchangeable.

Figure 2.2, p. 38, presents two images produced by the *Yohkoh* satellite, a joint effort of the National Astronomical Observatory of Japan, the University of Tokyo, Lockheed Palo Alto Research Laboratory, and NASA. Note the correlation between the sunspot grouping and an intense X-ray source on the Sun. There is a video tape sequence that also is available from these sponsors that dramatically illustrates the capability of this new satellite. See: L. Acton, *et al.*, "The Yohkoh Mission for High-Energy Solar Physics," *Science* 258, October 23, 1992: 618–25.

9. What is the sunspot cycle? At what stage in the cycle were we in 2007?

A regular cycle exists for sunspot occurrences, averaging 11 years from maximum peak to maximum peak; however, the cycle may vary from 7 to 17 years. In recent cycles, a solar minimum occurred in 1976, whereas a solar maximum took place in 1979, when over 100 sunspots were visible. Another minimum was reached in 1986, and an extremely active solar maximum occurred in 1990 with over 200 sunspots, 11 years from the previous maximum, in keeping with the average. A sunspot minimum occurred in 1997, an intense maximum in 2001, and an expected minimum 2006–2007 maintains the average.

10. Describe Earth's magnetosphere and its effects on the solar wind and the electromagnetic spectrum.

Earth's outer defense against the solar wind is the magnetosphere, which is a magnetic force field surrounding Earth, generated by dynamo-like motions within our planet. As the solar wind approaches Earth, the streams of charged particles are deflected by the magnetosphere and course along the magnetic field lines. The extreme northern and southern polar regions of the upper atmosphere are the points of entry for the solar wind stream.

11. Compare the filtering aspects of the atmosphere to the astronaut's spacesuit shown in Figure 2.3, specifically the solar wind.

The atmosphere and the spacesuit both act as buffers, protecting the inhabitants from extremes of temperature, harmful wavelengths of electromagnetic radiation, the streams of charged particles in the solar wind, and natural and human-caused space debris.

12. Describe the various segments of the electromagnetic spectrum, from shortest to longest wavelength. What wavelengths are mainly produced by the Sun? Which are principally radiated by Earth to space?

See Figures 2.6 and 2.7. All the radiant energy produced by the Sun is in the form of electromagnetic energy and, when placed in an ordered range, forms part of the electromagnetic spectrum. The Sun emits radiant energy composed of 8 percent ultraviolet, X-ray, and gamma ray wavelengths; 47 percent visible light wavelengths; and 45 percent infrared wavelengths. Wavelengths emitted from the Earth back to the Sun are of lower intensity and composed mostly of infrared wavelengths.

13. What is the solar constant? Why is it important to know?

The average value of insolation received at the thermopause (on a plane surface perpendicular to the Sun's rays) when Earth is at its average distance from the Sun. That value of the solar constant is 1372 watts per square meter (W/m2). A *watt* is equal to one joule (a unit of energy) per second and is the standard unit of power in the SI-metric system. (See Appendix C in *Elemental Geosystems* for more information on measurement conversions.) In nonmetric calorie heat units, the solar constant is expressed as approximately 2 calories (1.968) per cm2 per minute, or 2 langleys per minute (a langley being 1 cal per cm2). A *calorie* is the amount of energy required to raise the temperature of one gram of water (at 15°C) one degree Celsius and is equal to 4.184 joules.

Knowing the amount of insolation intercepted by Earth is important to climatologists and other scientists as a basis for atmospheric and surface energy measurements and calculations.

14. Select 0° or 90° latitude on Figure 2.10 and compare approximately the amount of energy received. What differences do you observe? On the map where do you find the largest net radiation? The lowest net radiation? See Figure 2.10.

Note the watts per m^2 received at each month for specific latitudes. Also see Figure 2.9 for relative insolation receipts by latitude.

15. If Earth were flat and oriented perpendicularly to incoming solar radiation (insolation), what would be the latitudinal distribution of solar energy at the top of the atmosphere?

The atmosphere is like a giant heat engine driven by differences in insolation from place to place. If Earth were flat there would be an even distribution of energy by latitude with no differences from place to place and therefore little motion produced.

_____

• *Define* solar altitude, solar declination, and daylength and *describe* the annual variability of each—Earth's seasonality.

16. Contrast the concept of seasonality between the equator and the polar regions?

The equatorial regions are lacking seasonality that the polar regions have. Along the equator, every day is pretty much like every other with regards to daylength, while at the poles there are only two days in each calendar year. At the poles the sun rises above the horizon for three months, then gradually slips toward the horizon for three months; this is then followed by the sun continuing to fall farther below the horizon for three months, and finally, the sun returns to the horizon for three months.

Notes on Calendars. A lot can be done with the calendar to highlight the concept of seasons. The need for an accurate calendar to schedule planting and other activities was recognized early as a necessity. As early as 2500 B.C. the Babylonians were able to measure the length of a year, complete with the concept of weeks and months, and to establish a calendar based on lunar cycles. The Egyptians had established a calendar by 423 B.C. that was based on daily and annual solar cycles. The Mayan calendar in the New World, based on a complex of interacting cycles, was extremely accurate and assigned specific names to every day, week, and month.

In 46 B.C., Julius Caesar established a modified version of the Egyptian effort. And, of course the names of some months were eventually thrown off by two months (December is the 12th month, September is the 9th month, etc.) because Julius and Augustus both added months in their own names—July and August. The Julian calendar, by B.C. 1582, through accumulated errors, differed by more than 10 days from what the Sun was doing in the sky. Consequently, on 4 October 1582, Pope Gregory XIII instituted a new calendar that would correct many of the flaws in the Julian calendar. He adjusted for past errors by declaring that for the year 1582, "October 4th shall be followed by October 15th." The Gregorian calendar was not adopted in Britain until 1752, Russia in 1919, and China in 1949. In this high-technology era of atomic clocks, corrections of seconds are made occasionally, with leap days normally taken every four years (29 February) so that clocks and calendars are kept in sync with the heavens. In the complex Gregorian mess, 2000 is a leap year. Remember, 2000 was not the end of the Millennium. 2000 was not the 20th hundred-year yet. In the goofy Gregorian calendar there is no year zero—1 B.C. went to A.D. 1. So, years 1 to 100 were the first century, and years 101 to 200 were the second century, and so on. Therefore, 1901 to 2000 is the 20th century and 2001 begins the 21st century.

17. The concept of seasonality refers to what specific phenomena? How do these two aspects of seasonality change during the year at 0° latitude? At 40°? At 90°?

Seasonality refers to both the seasonal variation of the Sun's rays above the horizon and changing daylengths during the year. Seasonal variations are a response to the change in the Sun's altitude, or the angular difference between the horizon and the Sun. Seasonality also means a changing duration of exposure, or daylength, which varies during the year depending on latitude. People living at the equator always receive equal hours of day and night, whereas people living along 40° N or S latitude experience about six hours' of difference in daylight hours between winter and summer; those at 50° N or S latitude experience almost eight hours of annual daylength variation. At the polar extremes, the range extends from a six-month period of no insolation to a six-month period of continuous 24-hour days.

18. Differentiate between the Sun's altitude and its declination at Earth's surface.

The Sun's altitude is the angular difference between the horizon and the Sun. The Sun is directly overhead at zenith only at the subsolar point. The Sun's declination—that is, the angular distance from the equator to the place where direct overhead insolation is received—annually migrates through 47 degrees of latitude between the two tropics at 23.5° N and 23.5° S latitudes.

19. For the latitude at which you live, how does daylength vary during the year? How does the Sun's altitude vary? Does your local newspaper publish a weather calendar containing such information?

Check out Figure 2.15 for some guidance.

Seasons and Health Issues. Imagine living in Barrow, Alaska, where the night lasts for 65 days from sunset on 19 November until sunrise at 1:09 P.M. local time 23 January. Scientists are studying what the effects of this are on an individual's biological rhythms, moods, and behavioral patterns. Add to this that the outside temperature is approximately −57°C (−70°F), so that people are indoors more in winter. The effects on people in Fairbanks, where night is about 21 hours in length, are being studied. Effects on those living in higher latitudes are being studied as well. Some actually develop a seasonal disorder related to these long periods without the Sun. Treatment involves "phototherapy," full-spectrum light exposure. This light simulates the Sun's spectrum at the surface at about the time of the onset of twilight. Whatever the results of such research, it appears that to stay active and busy is important, that women are affected more than men, that Caucasians are affected more than native Eskimos, and that adults are affected more than children—some 35 million people are possibly affected.

Refer to the following: Bruce Bower, "Here Comes the Sun," *Science News*, 142, July 25, 1992: 62–3. This reviews scientific efforts to understand winter depression known as "seasonally affective disorder," or SAD. Another source is Dr. Norman E. Rosenthal, National Institute of Mental Health, Bethesda, M.D.

20. List the five physical factors that operate together to produce seasons.

Table 2.1 details these factors: Earth's revolution and rotation, its tilt and fixed-axis orientation, and its sphericity.

21. Describe Earth's revolution and rotation, and differentiate between them.

The structure of Earth's orbit and revolution about the Sun is described in Figures 2.1 and 2.11. Earth's revolution determines the length of the year and the seasons. Earth's rotation, or turning, is a complex motion that averages 24 hours in duration. Rotation determines daylength, produces the apparent deflection of winds and ocean currents, and produces the twice daily action of the tides in relation to the gravitational pull of the Sun and the Moon. Earth's axis is an imaginary line extending through the planet from the North to South geographic poles.

22. Define Earth's present tilt relative to its orbit about the Sun.

Think of Earth's elliptical orbit about the Sun as a level plane, with half of the Sun and Earth above the plane and half below. This level surface is termed the plane of the ecliptic. Earth's axis is tilted 23.5° from a perpendicular to this plane.

• *Construct* a general model of the atmosphere based on the criteria composition, temperature, and function, and *diagram* this model in a simple sketch.

23. What is air? Where did the components in Earth's present atmosphere originate?

Earth's atmosphere is a unique reservoir of gases, the product of billions of years of development. The modern atmosphere is probably Earth's fourth general atmosphere. The principal substance of this atmosphere is air, the medium of life as well as a major industrial and chemical raw material. Air is a simple additive mixture of gases that is naturally odorless, colorless, tasteless, and formless, blended so thoroughly that it behaves as if it were a single gas.

The primordial, evolutionary, and living atmospheres are discussed in Chapter 2. The outgassing hypothesis is discussed in Chapter 5. This modern atmosphere is, in reality, a gaseous mixture of ancient origins—the sum of all the exhalations and inhalations of life on Earth throughout time.

24. In view of the analogy by Lewis Thomas, characterize the various functions the atmosphere performs that protect the surface environment.

A membrane around a cell regulates the interactions of the delicate inner workings of that cell and the potentially disruptive outer environment. Each membrane is very selective as to what it will and will not allow to pass. The modern atmosphere acts as Earth's protective membrane. The atmosphere absorbs and interacts with harmful wavelengths of electromagnetic radiation, the streams of charged particles in the solar wind, and natural and human-caused space debris—all protecting the delicate aspects of the biosphere in the lower troposphere. As critical as the atmosphere is to us, it represents only a thin-skinned envelope amounting to less than one-millionth of Earth's total mass.

The dramatic and poignant quote from Lewis Thomas' *Lives of the Cell,* comparing the presence and function of the atmosphere to a cell membrane, is a wonderful metaphor (Figure 2.17, p. 48). Dr. Thomas passed away in 1993 at the age of 81, active until the end and still alive in everyone who has read his works. Thomas said, "Statistically the probability of any one of us being here is so small that you'd think the mere fact of existing would keep us all in a contented dazzlement of surprise." Maybe through the perspective of physical geography we can raise students to a greater sense of wonder about our life-supporting Earth. George Will in a December 1993 *Washington Post* column said, "A quiet but insistent voice....He watched in amazement undiminished over the years."

25. What three distinct criteria are employed in dividing the atmosphere for study?

The atmosphere is conveniently classified using three criteria: composition, temperature, and function. Based on chemical *composition*, the atmosphere is divided into two broad regions: the heterosphere and the homosphere. Based on *temperature*, the atmosphere is divided into four distinct zones: the thermosphere, mesosphere, stratosphere, and troposphere. Finally, two specific zones are identified on the basis of *function* relative to their role of removing most of the harmful wavelengths of solar radiation—these are the ionosphere and the ozonosphere, or ozone layer. See the left-hand column in Figure 2.18.

26. Describe the overall temperature profile of the atmosphere and list the four layers defined by temperature.

The temperature profile in Figure 3.17 shows that temperatures rise sharply in the thermosphere, up to 1200°C (2200°F) and higher at the top of the atmosphere. Individual molecules of nitrogen and oxygen and atoms of oxygen are excited to high levels of vibration from the intense radiation present in this portion of the atmosphere. This *kinetic energy*, the energy of motion, is the vibrational energy stated as temperature, although the density of the molecules is so low that little heat is produced. Heating in the lower atmosphere closer to Earth differs because the active molecules in the denser atmosphere transmit their kinetic energy as *sensible heat*. The mesosphere is the coldest portion of the atmosphere, averaging –90°C (–130°F), although that temperature may vary by 25C° to 30C° (45F° to 54F°).

Temperatures increase throughout the stratosphere, from –57°C (–70°F) at 20 km (tropopause), warming to freezing at 50 km (stratopause). The tropopause is defined by an average temperature of –57°C (–70°F), but its exact location varies with the seasons of the year, latitude, and sea level temperatures and pressures.

Tropospheric temperatures decrease with increasing altitude at an average of 6.4C° per km (3.5F° per 1000 ft), a rate known as the normal lapse rate. The actual lapse rate at any particular time and place under local weather conditions is called the environmental lapse rate, and may vary greatly from the normal lapse rate. The lab manual and student study guide both have students work with these concepts.

27. Describe the two divisions of the atmosphere based on composition.

The homosphere comprises an even mixture of gases:

Symbol	% by Vol.	Parts per million
N_2	78.084	780,840
O_2	20.946	209,460
Ar	0.934	9,340
CO_2	0.038	380

The atmosphere is a vast reservoir of relatively inert nitrogen, principally originating from volcanic sources. In the soil, nitrogen is bound up by nitrogen-fixing bacteria and is returned to the atmosphere by denitrifying bacteria that remove it from organic materials. Oxygen, a by-product of photosynthesis, is essential for life processes. Along with other elements, oxygen forms compounds that compose about half of Earth's crust. Argon is a residue from the radioactive decay of a form of potassium (^{40}K). A slow process of accumulation accounts for all that is present in the modern atmosphere. Argon is completely inert and as a noble gas is unusable in life processes. Although it has increased over the past 200 years, carbon dioxide is included in this list of stable gases. It is a natural by-product of life processes, and the implications of its current increase are critical to society and the future. The role of carbon dioxide in the gradual warming of Earth is discussed in Chapters 3 and 6.

The heterosphere is arranged in layers based on the atomic weight of atoms and gases and reactions with the incoming solar beam.

28. What are the two primary functional layers of the atmosphere and what do each do?

Ionosphere and stratosphere. See Figure 2.23 and observe the shorter wavelengths that are filtered in the functional layers of the atmosphere.

———————

• *Describe* conditions within the stratosphere; specifically *review* the function and status of the ozonosphere (ozone layer).

29. Why is stratospheric ozone (O_3) so important? Describe the effects created by increases in ultraviolet light reaching the surface.

Ozone absorbs wavelengths of ultraviolet light and subsequently reradiates this energy at longer wavelengths as infrared energy. Through this process, most harmful ultraviolet radiation is converted, effectively "filtering" it and safeguarding life at Earth's surface. See Focus Study 2.1 for greater discussion of the effects caused by ozone depletion.

30. Summarize the ozone predicament and the present trends, and any treaties that intend to protect the ozone layer.

See Focus Study 2.1 for a review of current treaties and their status. Environment Canada has launched a scientific observatory to study the ozone layer over Canada. This high-Arctic facility now is operational at a remote weather station near Eureka on Ellesmere Island, N.W.T., about 1000 km from the North Pole. The Arctic ozone depletion affects concentrations at lower latitudes and thus should draw the United States into further cooperation.

31. Evaluate Crutzen, Rowland, and Molina's use of the scientific method in investigating stratospheric ozone depletion.

The possible depletion of the ozone layer by human activity was first suggested during the summer of 1974 by University of California, Irvine professors F. Sherwood Rowland and Mario J. Molina. Rowland and Molina hypothesized that the stable, large CFC molecules remained intact in the atmosphere, eventually migrating upward and working their way into the stratosphere. CFCs do not dissolve in water and do not break down biologically. The increased ultraviolet light encountered by the CFC molecule in the stratosphere dissociates, or splits, the CFC molecules, releasing chlorine (Cl) atoms and forming chlorine oxide (ClO) molecules. At present, after much scientific evidence and verification of actual depletion of stratospheric ozone by chlorine atoms (1987–1990), even industry has had to admit that the problem is real.

The scientific method is discussed in Focus Study 1.1. Recall in the chapter when Dr. Rowland stated his frustration: "What's the use of having developed a science well enough to make predictions, if in the end all we are willing to do is stand around and wait for them to come true....Unfortunately, this means that if there is a disaster in the making in the stratosphere, we are probably not going to avoid it." (Roger B. Barry, "The Annals of Chemistry," *The New Yorker*, 9 June 1986, p. 83.) The Nobel Committee awarded these men, and Paul Crutzen, the 1995 Nobel Prize for chemistry for their important discovery and actions (see page 55).

In 1994 an international scientific consensus confirmed previous assessments of the anthropogenic (human-caused) disruption of the ozone layer—in other words chlorine atoms and chlorine monoxide molecules in the stratosphere are of human origin. The report, *Scientific Assessment of Ozone Depletion*, was prepared by NASA, NOAA, United Nations Environment Programme, and World Meteorological Organization.

• *Distinguish* between natural and anthropogenic variable gases and materials in the lower atmosphere and describe the sources and effects of air pollution.

32. Why are anthropogenic gases more significant to human health than those produced from natural sources?

Natural sources produce a greater quantity of nitrogen oxides, carbon monoxide, and carbon dioxide than do certain human-made sources. However, any attempt to diminish the impact of human-made air pollution through a comparison with natural sources is irrelevant, for we have co-evolved and adapted to the presence of certain natural ingredients in the air. We have not evolved in relation to the concentrations of anthropogenic (human-caused) contaminants found in our metropolitan areas.

33. In what ways does a temperature inversion worsen an air pollution episode? Why?

A temperature inversion occurs when the normal decrease of temperature with increasing altitude reverses at any point from ground level up to several thousand feet. Such an inversion most often results from certain weather conditions, for example when the air near the ground is radiatively cooled on clear nights or when cold air drains into valleys. The warm air inversion in Figure 2.26 prevents the vertical mixing of pollutants with other atmospheric gases. Thus, instead of being carried away, the pollutants are trapped below the inversion layer. During the winter months in eastern and midwestern regions of the United States, high pressure areas created by subsiding cold air masses produce inversion conditions that trap air pollution. In the West, summer subtropical high pressure systems also cause inversions and produce air stagnation.

34. What is the difference between industrial smog and photochemical smog?

The air pollution associated with coal-burning industries is known as industrial smog. The term smog was coined by a London physician at the turn of this century to describe the combination of fog and smoke containing sulfur, an impurity found in fossil fuels. The combination of sulfur and moisture droplets forms a sulfuric acid mist that is extremely dangerous in high concentrations.

Photochemical smog is another type of pollution that was not generally experienced in the past but developed with the advent of the automobile. Today, it is the major component of anthropogenic air pollution. Photochemical smog results from the interaction of sunlight and the products of automobile exhaust. Although the term *smog* is a misnomer it is generally used to describe this phenomenon. Smog is responsible for the hazy appearance of the sky and the reduced intensity of sunlight in many of our cities.

35. Describe the relationship between automobiles and the production of ozone and PAN in city air. What are the principal negative impacts of these gases?

The nitrogen dioxide derived from automobiles, and power plants to a lesser extent, is highly reactive with ultraviolet light, which liberates atomic oxygen (O) and a nitric oxide (NO) molecule (Figure 2.28, p. 61). The free oxygen atom combines with an oxygen molecule (O_2) to form the oxidant ozone (O_3); the same gas that is beneficial in the stratosphere is an air pollution hazard at Earth's surface. In addition to forming O_3, the nitric oxide (NO) molecule reacts with hydrocarbons (HC) to produce a whole family of chemicals generally called peroxyacetyl nitrates (PAN). PAN produces no known health effects in humans but is particularly damaging to plants, which provided the clue for discovery of these photochemical reactions.

36. How are sulfur impurities in fossil fuels related to the formation of acid in the atmosphere and acid deposition on the land?

Certain anthropogenic gases are converted in the atmosphere into acids that are removed by wet and dry deposition processes. Nitrogen and sulfur oxides (NOx, and SOx) released in the combustion of fossil fuels can produce nitric acid (HNO_3) and sulfuric acid (H_2SO_4) in the atmosphere. Precipitation as acidic as pH 2.0 has fallen in the eastern United States, Scandinavia, and Europe. By comparison, vinegar and lemon juice register slightly less than 3.0. Aquatic life perishes when lakes drop below pH 4.8.

37. In summary, what are the results from the first 20 years under Clean Air Act regulations? In your opinion, do you see viable arguments for its repeal?

The Benefits of the Clean Air Act, 1970 to 1990 (Office of Policy, Planning, and Evaluation, U. S. EPA) calculated the following:

- The *total direct cost* to implement the Clean Air Act for all federal, state, and local rules from 1970 to 1990 was *$523 billion* (in 1990-value dollars). This cost was borne by businesses, consumers, and government entities.
- The estimate of *direct monetized benefits* from the Clean Air Act from 1970 to 1990 falls in a range from $5.6 to $49.4 trillion with a central mean of *$22.2 trillion.*
- Therefore, the *net financial benefit* of the Clean Air Act is *$21.7 trillion*! "The finding is overwhelming. The benefits far exceed the costs

of the CAA in the first 20 years," said Richard Morgenstern, associate administrator for policy planning and evaluation at the EPA.

The benefits to society, directly and indirectly, have been widespread across the entire population: improved health and environment, less lead to harm children, lowered cancer rates, less acid deposition, an estimated 206,000 fewer deaths related to air pollution related deaths in 1990 alone, among many benefits described. These benefits took place during a period in which the U. S. population grew by 22 percent and the economy expanded by 70 percent.

Overhead Transparencies for Chapter 2

	Chapter 2	Solar Energy, Seasons, and the Atmosphere
20.	2.1	Earth's orbit and the solar system and the galaxy
21.	2.3	Astronaut and solar wind experiment
22.	2.6	The electromagnetic spectrum of radiant energy
23.	2.7	Solar and terrestrial energy distribution/wavelengths
24.	2.8	Earth's energy budget simplified
25.	2.9	Insolation receipts and Earth's curved surface
26.	2.10	Daily net radiation patterns at top of the atmosphere
27.	2.11	Earth's revolution and rotation
28.	2.12	The plane of the Earth's orbit (the plane of the ecliptic)
29.	2.13	Annual march of the seasons
30.	2.15	Seasonal observations—sunrise, noon, sunset at 40° N latitude
31.	2.17 a, b	Profile of the modern atmosphere; Space Shuttle sunset
32.	2.18 a, b	Density decrease with altitude; Atmospheric pressure profile
33.	2.19	Composition of the homosphere
34.	2.20	Temperature profile of the troposphere
35.	2.21	The atmosphere protects Earth surface environment
36.	F.S. 2.1.1	Antarctic ozone hole, TOMS image, 10/1/2003
37.	F.S. 2.1.2 a, b	Chemical evidence of ozone damage by humans
38.	2.24 a, b, c	Normal and inversion temperature profiles
39.	2.26	Photochemical reactions
40.	2.28	Air pollution emission trends: 1970–2001
41.	F.S. 2.2-2	Sulfate wet distribution

3

Atmospheric Energy and Global Temperatures

Earth's biosphere pulses daily, weekly, and yearly with flows of energy. Think for a moment of the annual pace of your own life, your wardrobe, gardens, and lifestyle activities—all reflect shifting seasonal energy patterns. Thus begins the culmination of the passage of energy from the Sun, across space, to the top of the atmosphere, down through the layers of the atmosphere to Earth's surface. This entire process should be thought of as a vast flow-system with energy cascading through fluid Earth systems. The beginning lecture can mention this flow-system and the journey of the past chapter.

Air temperature has a remarkable influence upon our lives, both at the microlevel and at the macrolevel. A variety of temperature regimes worldwide affect entire lifestyles, cultures, decision-making, and resources spent. Global temperature patterns presently are changing in a warming trend that is affecting us all and is the subject of much scientific, geographic, and political interest. Our bodies sense temperature and subjectively judge comfort, reacting to changing temperatures with predictable responses.

This chapter presents principles and concepts that are synthesized on the January and July temperature maps, and on the annual range of temperature map. The chapter then relates these temperature patterns and concepts directly to the student with a discussion of apparent temperatures—the wind chill and heat index charts. Focus Study 3.2 introduces some essential temperature concepts to begin our study of world temperatures.

We take a look at how Earth's temperature system appears to be in a state of dynamic change as concerns about global warming and potential episodes of global cooling are discussed in Chapter 7. An overview of the latest scientific findings concerning global climate change is in that chapter.

Outline Headings and Key Terms

The first-, second-, and third-order headings that divide Chapter 3 serve as an outline for your notes and studies. The key terms and concepts that appear boldface in the text are listed here under their appropriate heading in bold italics. All these highlighted terms appear in the text glossary. Note the check-off box (❑) so you can mark your progress as you master each concept. Your students have this same outline in their *Student Study Guide*. The ✆ icon indicates that there is an accompanying animation or other resource on the CD.

The outline headings for Chapter 3:

Energy Essentials
✆ **Global Albedo Values**
✆ **Global Shortwave Radiation**
✆ **Global Net Radiation**
✆ **Global Latent Heat Flux Values**
✆ **Global Sensible Heat**
   Energy Pathways and Principles
        ❑ *transmission*
   Insolation Input
   Scattering (Diffuse Radiation)
        ❑ *scattering*
        ❑ *diffuse radiation*
   Refraction
        ❑ *refraction*
   Albedo and Reflection
        ❑ *albedo*
        ❑ *reflection*
   Clouds and Atmosphere's Albedo
        ❑ *cloud-albedo forcing*
        ❑ *cloud-greenhouse forcing*
   Absorption
        ❑ *absorption*
   Conduction, Convection, and Advection
        ❑ *conduction*

The URLs related to this chapter of *Elemental
Geosystems* can be found at
http://www.prenticehall.com/christopherson

Key Learning Concepts

**After reading the chapter and using this study
guide, the student should be able to:**

• *Identify* the pathways of solar energy through the
troposphere to Earth's surface: transmission, refraction,
albedo (reflectivity), scattering, diffuse radiation,
conduction, convection, and advection.
• *Describe* the greenhouse effect, and the patterns of
global net radiation and surface energy balances.
• *Review* the temperature concepts and temperature
controls that produce global temperature patterns.
• *Interpret* the pattern of Earth's temperatures for
January and July and annual temperature ranges.
• *Contrast* wind chill and heat index and *determine*
human response to these apparent temperature effects.
• *Portray* typical urban heat island conditions and
contrast the microclimatology of urban areas with that
of surrounding rural environments.

Annotated Chapter Review Questions

• *Identify* **the pathways of solar energy through the
troposphere to Earth's surface: transmission,
refraction, albedo (reflectivity), scattering, diffuse
radiation, conduction, convection, and advection.**

**1. Diagram a simple energy balance for the
troposphere. Label each shortwave and longwave
component and the directional aspects of related
flows.**
    The keys are Figures 3.1 and 3.10, which
illustrate the Earth-atmosphere energy balance, which is
referred to at the beginning of this section. This figure
is reproduced in outline form without labels in the
Student Study Guide. Students are instructed to fill in
information, add labels, and even color in the
illustrations as they move along through this discussion.
If you use a chalkboard, this is an opportunity to draw
the energy budget along with the students as you
lecture. The illustration is also included in the overhead
transparency packet with this chapter.

2. Define refraction. How is it related to daylength? To a rainbow? To the beautiful colors of a sunset?

When insolation enters the atmosphere, it passes from one medium to another (from virtually empty space to atmospheric gas) and is subject to a bending action called *refraction*. In the same way, a crystal or prism refracts light passing through it, bending different wavelengths to different degrees, separating the light into its component colors to display the spectrum.

A rainbow (Figure 3.3) is created when visible light passes through myriad raindrops and is refracted and reflected toward the observer at a precise angle. Another example of refraction is a *mirage*, an image that appears near the horizon where light waves are refracted by layers of air of differing temperatures (densities) on a hot day.

An interesting function of refraction is that it adds approximately eight minutes of daylight for us. The Sun's image is refracted in its passage from space through the atmosphere, and so, at sunrise, we see the Sun about four minutes before it actually peeks over the horizon. Similarly, at sunset, the Sun actually sets but its image is refracted from over the horizon for about four minutes afterward. To this day, modern science cannot predict the exact time of sunrise or sunset within these four minutes, because the degree of refraction continually varies with temperature, moisture, and pollutants.

Notes on Sky Color. The principle known as Rayleigh scattering—named for English physicist Lord Rayleigh, who stated the principle in 1881—relates wavelength to the size of molecules or particles that cause the scattering. The general rule is the shorter the wavelength, the greater the scattering, and the longer the wavelength, the less the scattering. Shorter wavelengths of light are scattered by small gas molecules in the air. Thus, the shorter wavelengths of visible light, the blues and violets, are scattered the most and dominate the lower atmosphere. And because there are more blue than violet wavelengths in sunlight, a blue sky prevails. As the atmosphere thins with altitude, there are fewer molecules to scatter these shorter wavelengths, and the sky darkens. At 50 km, even though it may be daylight, the sky appears as at night and the stars become visible along with the Sun.

3. List several types of surfaces and their albedo values. Explain the differences among these surfaces. What determines the reflectivity of a surface?

A portion of arriving energy bounces directly back to space without being converted into heat or performing any work. This returned energy is called reflection, and it applies to both visible and ultraviolet light. The reflective quality of a surface is its albedo, or the relationship of reflected to incoming insolation expressed as a percentage (In Figure 3.6). In terms of visible wavelengths, darker colors have lower albedos, and lighter colors have higher albedos. On water surfaces, the angle of the solar beam also affects albedo values; lower angles produce a greater reflection than do higher angles. In addition, smooth surfaces increase albedo, whereas rougher surfaces reduce it. See Figure 3.6 for specific values.

4. Define the concepts transmission, absorption, diffuse radiation, conduction, and convection.

See the Glossary section in the text and Figure 3.7 for illustration.

• *Describe* **the greenhouse effect, and the patterns of global net radiation and surface energy balances.**

5. What are the similarities and differences between an actual greenhouse and the gaseous atmospheric greenhouse? Why is Earth's greenhouse changing?

In the greenhouse analogy, the glass is transparent to shortwave insolation, allowing light to pass through to the soil, plants, and wood planks inside. The absorbed energy is then radiated as infrared energy back toward the glass, but the glass effectively traps the longer infrared wavelengths and warms the air inside the greenhouse. Thus, the glass allows the light in but does not allow the heated air out.

In the atmosphere, the greenhouse analogy is not fully applicable because infrared radiation is not trapped as it is in a greenhouse. Rather, its passage to space is delayed as the heat is radiated and reradiated back and forth between Earth's surface and certain gases and particulates in the atmosphere. The present warming is associated with an increase in radiatively active greenhouse gases.

This discussion is important to portions of the rest of the text relative to the greenhouse effect and global climate change. The operation of Earth's greenhouse sets the stage for the discussion of future temperature trends and global warming, and future climate patterns and consequences of climatic warming in Chapter 6, sea level changes in Chapter 13, the Antarctic ice sheet in Chapter 14, ecosystem stability and climate change in Chapter 16, and the summary overview comments in Chapter 17. The text author chose to present global change in this manner instead of in an isolated focus study because this present trend is so spatially pervasive through many of Earth's systems. We hope this integrated approach will assist the student in seeing the complex interconnections that link all Earth systems and the human population.

6. Generalize the pattern of global net radiation. How might this pattern drive the atmospheric weather machine? (See Figure 3.11.)

In the equatorial zone, surpluses of energy dominate, for in those areas more energy is received than is lost. Sun angles there are high, with consistent daylength. However, deficits exist in the polar regions, where more energy is lost than gained. At the poles the Sun is extremely low in the sky, surfaces are light and reflective, and for six months during the year no insolation is received. This imbalance of net radiation from the equator to the poles drives the vast global circulation of energy and mass.

Figure 3.11 summarizes the latitudinal distribution of net radiation. This completes a full cycle of discussion that began in Chapter 2 with energy measurements at the top of the atmosphere. This is the net radiation portrait of the entire Earth-atmosphere system. A useful analogy for the students could be some version of the following:

Imagine that you are dealing with money instead of energy, so that at the equator there is the Tropical Branch Bank, and at the pole the Polar Branch Bank. You are in charge of accounts at both of these bank branches, and for reasons unknown to you, more deposits are made at the Tropical Branch than you are withdrawing. At the Polar Branch, you make more withdrawals than you do deposits. Fix in your mind that both accounts act as open-flow systems, with yourself in charge of the inputs and outputs. The resultant dollar amounts in your two accounts pose an interesting problem: one is full of surplus cash while the other account is in a deficit, with checks bouncing. Your solution? To balance your financial situation, you must transfer excess deposits from the surplus account to the deficit account. Of course this is what happens in the Earth-atmosphere system, with the transfer of energy and mass (water and water vapor).

7. In terms of surface energy balance, explain the term net radiation (NET R).

Net radiation is the net all-wave radiation available at Earth's surface; it is the final outcome of the entire radiation balance process discussed in this chapter. Net radiation (NET R) is the balance of all radiation, shortwave (SW) and longwave (LW), at Earth's surface.

As students travel about during the day (perhaps not as far afield as these sample stations), they can note the different surfaces and imagine each of the energy balance components. Ask them to consider the pathways for the expenditure of net radiation available at each surface: turbulent transfer, latent heat of evaporation, photosynthesis, conduction into the soil, or conduction and convection processes in bodies of water. And consider various alterations to those surfaces: clearing, paving, reforestation, and tilling. The students might speculate about what changes take place in the net radiation balance because of these alterations.

As a contrasting example, work through the equation as if we were applying it to the lunar surface. The NET R at the surface of the Moon is totally expended for G. The surface increases in temperature to almost the boiling point of water and decreasing in temperature an equal amount below freezing in shadow or at night. An astronaut in a spacesuit will receive total G on his sun-facing side and total loss of G on his shady side as he stands in the direct sunlight.

8. What are the expenditure pathways for surface net radiation? What kind of work is accomplished?

Output paths at the surface for the principal expenditures of net radiation from a nonvegetated surface include H (turbulent sensible heat transfer), LE (latent heat of evaporation), and G (ground heating and cooling). See bulleted items in the text chapter.

9. What is the role played by latent heat of evaporation in surface energy budgets?

Latent heat refers to heat energy that becomes stored in water vapor as water evaporates. Large quantities of latent heat are absorbed into water vapor during its change of state from liquid to gas. Conversely, this heat is released in its change of state back to a liquid (see Chapter 5). Because the evaporation of water is the principal method of transferring and dissipating heat surpluses vertically into the atmosphere, latent heat is the *dominant expenditure* of Earth's entire NET R. Latent heat links Earth's energy and water (hydrologic) systems and for most landscapes is the key component in surface energy budgets.

10. Compare the daily surface energy balances of El Mirage, California, and Pitt Meadows, British Columbia. Explain the differences.

El Mirage, California, at 35° N, is a hot desert location characterized by bare, dry soil with very little vegetation (Figure 3.14a). The summer day selected was clear, with a light wind in the late afternoon. El Mirage has little or no expenditure of energy for LE. With an absence of water and plants, most of the available radiant energy is dissipated as turbulent sensible heat, warming air and soil to high temperatures. The G component is higher in the morning, when winds are light and turbulent transfers are reduced.

The NET R at this desert location is quite similar to that of midlatitude, vegetated, and moist Pitt Meadows, British Columbia (Figure 3.14b). The energy balance data for Pitt Meadows, at 49° N, are plotted for a cloudless summer day. The Pitt Meadows landscape is able to retain much more of its energy because of a lower albedo (less reflection), the presence of more water and plants, and lower surface temperatures. The higher LE values are attributable to the moist environment of rye grass and irrigated mixed-orchard ground cover for the sample area, contributing to the more moderate sensible heat levels throughout the day.

11. Why is there a temperature lag between the highest Sun altitude and the warmest time of day? Relate your answer to insolation and temperature patterns during the day.

Incoming energy arrives throughout the illuminated part of the day, beginning at sunrise, peaking at local noon, and ending at sunset. As long as the incoming energy exceeds the outgoing energy, temperature continues to increase during the day, not peaking until the incoming energy begins to diminish in the afternoon as the Sun loses altitude. The warmest time of day occurs not at the moment of maximum insolation but at that moment when a maximum of insolation is absorbed. Thus, this temperature lag places the warmest time of day three to four hours after solar noon as absorbed heat is supplied to the atmosphere from the ground. Then, as the insolation input decreases toward sunset, the amount of heat lost exceeds the input, and temperatures begin to drop until the surface has radiated away the maximum amount of energy, just at dawn.

The annual pattern of insolation and temperature exhibits a similar lag. For the Northern Hemisphere, January is usually the coldest month, occurring after the winter solstice, the shortest day in December. Similarly, the warmest months of July and August occur after the summer solstice, the longest day in June.

• *Review* the temperature concepts and temperature controls that produce global temperature patterns.

12. Explain the effect of altitude on air temperature. Why is air at higher altitudes lower in temperature? Why does it feel cooler standing in shadows at higher altitudes than at lower altitudes?

Air temperatures in the troposphere decrease with increasing elevation above Earth's surface (recall that the normal lapse rate of temperature change with altitude is 6.4°C/1000 m or 3.5°F/1000 ft). Thus, worldwide, mountainous areas experience lower temperatures than do regions nearer sea level, even at similar latitudes. Temperatures may decrease noticeably in the shadows and shortly after sunset. Surfaces both heat rapidly and lose their heat rapidly at higher altitudes. This is a result of lower density air having a lower specific heat, because a given volume of air can hold less heat energy than a denser volume of air.

13. What noticeable effect does air density have on the absorption and radiation of energy? What role does altitude play in that process?

The density of the atmosphere also diminishes with increasing altitude, as discussed in Chapter 2. As the atmosphere thins, its ability to absorb and radiate heat is reduced. The consequences are that average air temperatures at higher elevations are lower, nighttime cooling increases, and the temperature range between day and night and between areas of sunlight and shadow also increases.

14. How is it possible to grow moderate-climate-type crops such as wheat, barley, and potatoes at an elevation of 4103 m (13,460 ft) near La Paz, Bolivia, so near the equator?

The combination of elevation and low-latitude location guarantees La Paz nearly constant daylength and moderate temperatures, averaging about 9°C (48°F) for every month. Such moderate temperature and moisture conditions lead to the formation of more fertile soils than those found in the warmer, wetter climate of Concepción.

15. Describe the effect of cloud cover with regard to Earth's temperature patterns —review cloud-albedo forcing and cloud-greenhouse forcing.

Clouds are moderating influences on temperature, producing lower daily maximums and higher nighttime minimums. Acting as insulation, clouds hold heat energy below them at night, preventing more rapid radiative losses, whereas during the day, clouds reflect insolation as a result of their high albedo values. The moisture in clouds both absorbs and liberates large amounts of heat energy, yet another factor in moderating temperatures at the surface.

• *Interpret* the pattern of Earth's temperatures for January and July and annual temperature ranges.

16. What is the thermal equator? Describe its location in January and in July. Explain why it shifts position annually.

The thermal equator, a line connecting all points of highest mean temperature (black dashed line on the maps in Figure 3.24 and 3.26), in January trends southward into the interior of South America and Africa, indicating higher temperatures over landmasses. The thermal equator shifts northward in July with the high summer Sun and reaches the Persian Gulf-Pakistan-Iran area. (relate this to specific heat).

17. Observe trends in the pattern of isolines over North America and compare the January average temperature map with the July map. Why do the patterns shift locations?

Isotherms over North America vary with seasonal shifts in the Sun's declination and daylength. As the ITCZ shifts toward the Southern Hemisphere, North America receives less solar radiation, due to reduced daylength and a lower angle of incidence this time of year. This correlates to reduced temperatures experienced during January in North America.

As the ITCZ shifts toward the Northern Hemisphere, North America receives more solar radiation, due to increased daylength and a greater angle of incidence. Temperatures in North America will be more extreme. Due to the specific heat properties of land, North America will experience greater temperature extremes, the continent will lose energy rapidly in January due to land's low heat capacity, and the continent will heat rapidly in July due to the low amount of energy that is required to heat land.

18. Describe and explain the extreme temperature range experienced in north-central Siberia between January and July.

As you might expect, the largest temperature ranges occur in subpolar locations in North America and Asia, where average ranges of 64°C (115°F) are recorded. The Verkhoyansk region of Siberia is probably the greatest example of continentality on Earth. The coldest area on the map is in northeastern Siberia in Russia. The cold experienced there relates to consistent clear, dry air, small insolation input, and an inland location far from any moderating maritime effects. Verkhoyansk and Omakon, Siberia, Russia, each have experienced a minimum temperature of −68°C (−90°F) and a daily average of −50.5°C (−58.9°F) in January. Verkhoyansk experiences at least seven months of temperatures below freezing, including at least four months below −34°C (−30°F)! July temperatures in Verkhoyansk average more than 13°C (56°F), which represents a 63C° (113F°) seasonal variation between winter and summer averages.

You probably have your own favorite sets of stations to portray continental and marine characteristics. I present San Francisco and Wichita, Vancouver and Winnipeg, and Trondheim and Verkhoyansk and have included three pairs of cities in the overhead transparency packet. In simplest terms, these demonstrate continentality of the interior stations and the greater moderation of temperature characteristics of the coastal stations. With a more sophisticated analysis, cities in the western Basin and Range Province such as Elko, Nevada, probably exhibit the least maritime influence in the United States measured by indicator formulas, (i.e., the greatest degree of continentality).

19. Where are the hottest places on Earth? Are they near the equator or elsewhere? Explain. Where is the coldest place on Earth?

The hottest places on Earth occur in Northern Hemisphere deserts during July. These deserts are areas of clear and dry skies and strong surface heating, with virtually no surface water and few plants. Locations such as portions of the Sonoran Desert area of North America and the Sahara of Africa are prime examples. Africa has recorded shade temperatures in excess of 58°C (136°F), such as a record set on 13 September 1922 at Al-Aziziyah, Libya (32° 32' N; 112 m or 367 ft elevation). The highest maximum and annual average temperatures in North America occurred in Death Valley, California, where the Greenland Ranch Station (37° N; −54.3 m or −178 ft below sea level) reached 57°C (134°F) in 1913.

July is a time of 24-hour-long nights in Antarctica. The lowest natural temperature reported on Earth occurred on 21 July 1983 at the Russian research base at Vostok, Antarctica (78°27 S, elevation 3420 m or 11,220 ft): a frigid −89.2°C (−128.56°F). For comparison, such a temperature is 11°C (19.8°F) colder than dry ice (solid carbon dioxide)!

Many sources of climatic data are available, including those on the Internet and accessed through our Home Page. A few print sources are suggested below:

The National Weather Service maintains the *Climatology of the United States* for each of the 50 states and *World Weather Records*, updated in 1979. Local weather service offices prepare reports for metropolitan areas. One example is Tony Martini's "Climate of Sacramento, California," NOAA Technical Memorandum NWS WR-65 Sacramento: Weather Service Office, April 1990, 70 pp. Check to see if your local or state climatologist has prepared such a report. For Canada, contact the Atmospheric Environment Service, *Climatic Normals* publications.

Landsberg, Helmut E., ed. *World Survey of Climatology*, 15 volumes published 1969–1984. New York: Elsevier, North Holland.

Pearce, E.A. and C.G. Smith. *The World Weather Guide*. London: Hutchinson and Company, 1984.

Riordan, Pauline and Paul G. Bourget. *World Weather Extremes*. Fort Belvoir, VA: U.S. Army Corp of Engineers, 1985, available from USGPO.

Rudloff, Willy. *World Climates*. Stuttgart, Germany: Wissenschaftliche Verlagsgesellschaft, 1981.

Ruffner, James A. and Frank E. Blair, eds. *The Weather Almanac*. New York: Avon Books, 1977.

Wernstedt, Frederick L. *World Climatic Data*. Lemont, PA: Climatic Data Press, 1972.

Willmott, Cort J., John R. Mather, and Clinton M. Rowe. *Average Monthly and Annual Surface Air Temperature and Precipitation Data for the World*. Part 1, "The Eastern Hemisphere," and Part 2, "The Western Hemisphere." Elmer, NJ: C. W. Thornthwaite Associates and the University of Delaware, 1981.

For a complete survey of CD-ROM resources see: Clifford F. Mass, "The Application of Compact Discs (CD-ROM) in the Atmospheric Sciences and Related Fields: An Update," *Bulletin of the American Meteorological Society*, 74 no. 10, October 1993: 1901–1908. The article reviews CD-ROM technology and has two appendices: "Currently Available CD-ROM Titles in the Atmospheric Sciences and Related Disciplines," and "Contact Information for CD-ROM Vendors." The article references several data sets for climatological information.

• *Contrast* wind chill and heat index and *determine* human response to these apparent temperature effects.

20. What is the wind chill temperature on a day with an air temperature of –12°C (10°F) and a wind speed of 32 kmph (20 mph)?

32 kmph (20 mph) of wind with a temperature of -12°C (10°F) results in a wind chill reading of -23°C (-9°F).

21. On a day when temperature reaches 37.8°C (100°F), how does a relative humidity reading of 50% affect apparent temperature?

37.8°C (100°F), with a relative humidity reading of 50% results in a heat index of category II (very hot) and an apparent temperature of 49°C (120°F).

22. What is the basis for the urban heat-island concept? Describe the climatic effects attributable to urban as compared with nonurban environments. What did NASA determine from the UHIPP overflight of Sacramento (Figure 3.30)?

The surface energy characteristics of urban areas possess unique properties similar to energy balance traits of desert locations. See the six items of analysis in the chapter and the contents of Figure 3.28, 3.29, and 3.30.

23. Have you experienced any condition graphed in Focus Study 3.2, Figures 3.2.1 or 3.2.2? Explain.

Personal analysis and response.

24. Assess the potential for solar energy applications in our society. What are some negatives? What are some positives?

Solar energy systems can generate heat energy of an appropriate scale for approximately half the present population in the United States (space heating and water heating). In marginal climates, solar-assisted water and space heating is feasible as a backup; even in New England and the Northern Plains, solar-efficient collection systems prove effective. Kramer Junction, California, about 140 miles northeast of Los Angeles, has the world's largest operating solar electric-generating facility. The facility converts 23% of the sunlight it receives into electricity during peak hours. Rooftop photovoltaic electrical generation is now cheaper than power line construction to rural sites. Obvious drawbacks of both solar-heating and solar electric systems are periods of cloudiness and night, which inhibit operation.

The success of solar energy appears tied to the political arena, for without tax incentives and formal encouragement, in amounts at least equal to those given the fossil-fuel industry, it is difficult to operate such a plant.

I recommend that you write to the nonprofit organization listed below for their literature and designs. We purchased a simple box-cooker kit and find that it does extremely well for something made out of cardboard, one pane of glass, foil, and a reflector lid. The insulation in the walls is provided by crumpled newspaper and all the other materials are from recycled sources. You need about 15 minutes of direct sunlight per hour to cook food. We have cooked everything from pasta to bread to a whole turkey during the seven month period (April to October) at 40° N latitude. The only problem we have had was during the solar eclipse on July 11, 1991. With 59% of the Sun blocked by the Moon in Sacramento, California, temperature in the solar cooker dropped by 25°C (45°F).

For a copy of "Your Own Solar Box—How to Make and Use," "Teachers Guide Fun with the Sun," and "Leaders Guide, Spreading Solar Cooking," (all three for $12.) contact the following nonprofit group: Solar Cookers International (SCI), 1724 11th Street, Sacramento, California 95814; 916-444-6616; FAX-916-444-5379. The World Conference on Solar Cooker Use and Technology is held each year. A major effort is underway to place solar box cookers throughout the Third World. SCI also has videotapes, other teaching tools, and a newsletter dealing with solar cooking. Donations are tax deductible.

Relative to solar energy and the focus study with this chapter, there are some key generalizations that help organize energy resources for analysis. A first step is to analyze the actual end-use energy demand in the United States, or any country. The pattern of consumer demand most accurately dictates what methods and modes should be selected to supply and meet that energy need. Presently in the United States, 58% of end-use energy need is for heat roughly split one-half above and one-half below the boiling temperature of water. Another 38% is for mechanical motion, including 4% for electric motors. And finally, 4% is classified as necessary electrical. If the taxpayer is worried about taxes and big government, and is basically conservative, then having such a localized consumer-driven energy plan would seem wisest and most conservative of capital, political power, and economic control.

Secondly, let's briefly examine two categories of energy resources and their characteristics. The essays in the text on wind and solar resources should be viewed in the context of these two sets of energy characteristics. These two paradigms can form the basis of a class discussion if there is time.

Energy Resource Characteristics	
Centralized energy	**Decentralized energy**
Nonrenewable sources	Renewable sources
Indirect to the consumer	Direct to the consumer
Capital intensive	Labor intensive
Limited domestic supply	Unlimited domestic supply
Large foreign imports	No foreign imports
Monopolies/cartels	Widely available
"Big" government control	Home and neighborhood control
Concentrated tax advantages	Diffused tax advantages

Overhead Transparencies for Chapter 3

	Chapter 3	**Atmospheric Energy and Global Temperatures**
42.	3.1	Energy gained and lost by Earth's surface and atmosphere
43.	3.2	Insolation at Earth's surface
44.	3.4	Sun refraction photo and art
45.	3.5	Albedo values for various surfaces
46.	3.6 a, b; 3.8 a and b	Effects of clouds SW and LW (top); energy effects of two cloud types (bottom)
47.	3.8c and CO3	Jet contrails (top); cloud development from contrail
48.	3.7	Heat energy transfer process
49.	3.9 a, b	CERES shortwave (top) and longwave (bottom) global maps
50.	3.10	Detail of Earth-atmosphere energy balance
51.	3.11	Energy budget by latitude
52.	3.12	Daily radiation curves
53.	3.13	Surface energy budget
54.	3.15	Temperature scales (K, °C, °F)
55.	3.17	Temperature pattern variation by latitude
56.	3.18	Effects of altitude and latitude (La Paz and Concepción, Bolivia)
57.	3.20	Land-water heating differences
58.	3.22 a, b, scale	Sea-surface temperatures Feb. 1999 (top); SSTs July 1999 (bottom)
59.	3.23	Comparison: San Francisco and Wichita
60.	3.24 a,b,c	Global temperatures for January
61.	3.25	Comparison: Trondheim and Verkhoyansk
62.	3.26	Global temperatures for July
63.	3.28	Global annual temperature ranges
64.	3.29 and 3.30a	Urban environment (top); Typical urban heat island profile (bottom)
65.	H.L.C. 3.1.1	Loss of Arctic Ocean ice pack 1979, 2003
66.	F.S. 3.2.1 and 3.2.2	Wind-chill temperature index (top); Heat index (bottom)

Atmospheric and Oceanic Circulations

Earth's atmospheric circulation is an important transfer mechanism for both energy and mass. In the process, the energy imbalance between equatorial surpluses and polar deficits is partly resolved, Earth's weather patterns are generated, and ocean currents are produced. Human-caused pollution also is spread worldwide by this circulation, far from its points of origin. In this chapter we examine the dynamic circulation of Earth's atmosphere that carried Tambora's debris, Chernobyl's fallout, and Mount Pinatubo's ashen haze worldwide and carries the everyday ingredients oxygen, carbon dioxide, and water vapor around the globe. The subject of air pressure and its measurement, the pressure gradient, Coriolis, and friction forces are presented. We also consider Earth's wind-driven oceanic currents.

The keys to this chapter are in several integrated figures: the portrayal of winds by the *SEASAT* image (Figure 4.6), the three forces interacting to produce surface wind patterns and winds aloft (Figure 4.8), and the buildup through (Figure 4.13) to produce the idealized global circulation model.

Outline Headings and Key Terms

The first-, second-, and third-order headings that divide Chapter 4 serve as an outline for your notes and studies. The key terms and concepts that appear **boldface** in the text are listed here under their appropriate heading in ***bold italics***. All these highlighted terms appear in the text glossary. Note the check-off box (❑) so you can mark your progress as you master each concept. Your students have this same outline in their *Student Study Guide*. The ✪ icon indicates that there is an accompanying animation or other resource on the CD.

The outline headings for Chapter 4:

Wind Essentials
Air Pressure and Its Measurement
  ❑ ***air pressure***
  ❑ ***mercury barometer***
  ❑ ***aneroid baromete***r
Wind: Description and Measurement
  ❑ ***wind***
  ❑ ***anemometer***
  ❑ ***wind vane***
Global Winds
Driving Forces Within the Atmosphere
✪ **Wind Pattern Development**
✪ **Coriolis Force**
  ❑ ***pressure gradient force***
  ❑ ***Coriolis force***
  ❑ ***friction force***
Pressure Gradient Force
  ❑ ***isobars***
Coriolis Force
  ❑ ***geostrophic winds***
Friction Force
  ❑ ***anticyclone***
  ❑ ***cyclone***
Atmospheric Patterns of Motion
✪ **Global Patterns of Pressure**
✪ **Global Sea Level Pressure**
✪ **Global Wind Circulation, Hadley Cells**
✪ **Global Infrared**
✪ **Cyclones and Anticyclones**
✪ **Jet Stream, Rossby waves**
Primary High-Pressure and Low-Pressure Areas
  ❑ ***equatorial low-pressure trough***
  ❑ ***polar high-pressure cells***
  ❑ ***subtropical high-pressure cells***
  ❑ ***subpolar low-pressure cells***
Equatorial Low-Pressure Trough: ITCZ, Clouds and Rain
  ❑ ***intertropical convergence zone (ITCZ)***

The URLs related to this chapter of *Elemental
Geosystems* can be found at
http://www.prenticehall.com/christopherson

Key Learning Concepts

    After reading the chapter and using this
study guide, the student should be able to:

• *Define* the concept of air pressure and *portray* the
pattern of global pressure systems on isobaric maps
• *Define* wind and *describe* how wind is measured, how
wind direction is determined, and how winds are
named.
• *Explain* the four driving forces within the
atmosphere—gravity, pressure gradient force, Coriolis
force, and friction force—and *describe* the primary
high- and low-pressure areas and principal winds.
• *Overview* several multiyear oscillations of air
temperature, air pressure, and circulation in the Arctic,
Atlantic, and Pacific Oceans.
• *Describe* upper-air circulation and its support role for
surface systems and *define* the jet streams.
• *Explain* several types of local winds: land-sea breezes,
mountain-valley breezes, katabatic winds, and the
regional monsoons.
• *Describe* the basic pattern of Earth's major surface
and deep ocean currents.

Annotated Chapter Review Questions

• *Define* **the concept of air pressure and** *portray*
**patterns of global pressure systems on isobaric
maps.**

**1. How does air exert pressure? Describe the basic
instrument used to measure air pressure. Compare
the two different types of instruments discussed.**

    Air molecules—through their motion, size,
and number—produce pressure that is exerted on all
surfaces in contact with the air. The weight of the
atmosphere, or air pressure, crushes in on all of us;
fortunately, that same pressure is also inside us, pushing
out. The atmosphere exerts an average force of 14.7
pounds per square inch (approximately 1 kg/cm^2) at sea
level. Under the acceleration of gravity, air is
compressed and is denser near Earth's surface, rapidly
thinning out with increased altitude (Figure 2.18). This
decrease is measurable since air exerts its weight as a
pressure.

    Any instrument that measures air pressure is
called a barometer. Torricelli developed a mercury
barometer and established the average height of the
column of mercury in the barometric tube. The column
of mercury was counterbalanced by the mass of
surrounding air exerting an equivalent pressure on the
mercury in the vessel. A more compact design that
works without a meter tube of mercury is called an
aneroid barometer (Figure 4.2).

**2. What is normal sea level pressure in millimeters?
Millibars? Inches? Kilopascals?**

Using such instruments, normal sea level pressure is expressed as 1013.2 millibars (a way of expressing force per square meter of surface area), or 29.92 in., or 760 mm, of mercury; or 101.32 kPa. (10 mb = 1 kPa.)

3. What is a possible explanation for the beautiful sunrises and sunsets during the summer of 1992 in North America? Relate your answer to global circulation.

Following the eruption of Mount Pinatubo in June 1991, global atmospheric effects were noticed, although less in scale, similar in many ways to the greatest eruption in historic times. Early in April 1815, on an island named Sumbawa in present-day Indonesia, the volcano Tambora erupted violently, releasing an estimated 150 km^3 (36 mi^3) of material, an 80-times greater volume than that produced by the 1980 Mount Saint Helens eruption in Washington State. Some materials from Tambora—the aerosols and acid mists—were carried worldwide by global atmospheric circulation, creating a stratospheric dust veil. The result was both a higher atmospheric albedo and absorption of energy by the particulate materials injected into the stratosphere. Remarkable optical and meteorological phenomena resulting from the spreading dust were noted for months and years after the eruption. Beautifully colored sunsets and periods of twilight were enjoyed in London, Paris, New York, and elsewhere. In the summer of 1816, one year later, farmers in New England and Europe were shocked to experience frosts every month, and in some places, every week. Mean temperatures in the Northern Hemisphere apparently had dropped by 0.4-0.7°C (0.72-1.26°F). Although Tambora's eruption has not been conclusively tied to the cold summer in 1816, the impact of the eruption appears so large that, despite measurement uncertainties, it remains as powerful evidence of global circulation.

In Earth systems science, Earth is viewed ecologically, that is, in a holistic sense. We see overlapping and integrated physical, chemical, and biological systems. Geosystems does this using many threads that link content. This important volcanic eruption provides such content linkage.

4. Explain this statement, "The atmosphere socializes humanity, making all the world a spatially linked society." Illustrate your answer with some examples.

In the introduction to this chapter the text author gives several dramatic examples of the impact of global circulation: the Tambora eruption of 1815 and the Chernobyl nuclear disaster in 1986. The eruption of Mount Pinatubo that began in June 1991 is providing further evidence of this dynamic circulation. This eruption exceeded any other this century, including Mount Katmai, Alaska, which extruded 12 km^3 (2.88 mi^3) of pyroclastics. Also remember beyond these dramatic examples, global circulation mixes oxygen from principal production areas, presses against sails and kites, and guides and drives Earth's weather machine. See an initial mention by Richard A. Kerr, "Huge Eruption May Cool the Globe," in *Science* Vol. 252, No. 5014 (28 June 1991): p. 1780, and the new references under suggested readings at the end of the chapter. You no doubt noticed the afterglow (past sunset) during the fall of 1991 resulting from high-altitude ash and mist—more intense at lower latitudes than higher latitudes. Review the opening discussion of the spatial implications of this eruption in the introduction to this manual.

Examples abound for use in this introductory lecture. As an example, a U.S. satellite, with a plutonium-238 power plant on board, failed to achieve orbit following an April 1964 launch. The satellite burned on reentry, spreading minute amounts of radiation to monitoring stations throughout the Northern Hemisphere. Or in 1976, an above-ground nuclear test at the Lop Nor Chinese Test site sent radioactive debris into the atmosphere. Some 10 days later, fallout occurred over the United States in measurable quantities.

The Tambora eruption is unequaled in historic times for its output of ash and sulfur compounds. The volcano did trigger a global signature—this is undeniable. A direct connection with the cool summer of 1816 is still elusive and unproven despite several articles and books describing the effects. A couple of sources to consult:

Stothers, Richard B., "The Great Tambora Eruption in 1815 and Its Aftermath," *Science* Vol. 224, No. 4654 (15 June 1984): 1191–1198.

Stommel, Henry and Elizabeth Stommel, "The Year Without a Summer," *Scientific American* (June 1979): 176–186; also, by the same authors in *Volcano Weather—The Story of the Year Without a Summer*, Newport: Seven Seas Press, 1983.

With weather-related crop failures, the price of wheat soared in the years after the eruption to a level not reached again until 1972 and the Soviet grain shortage!

The aftermath of the Chernobyl accident continues to unfold more than a decade and a half later. Some 200,000 people in Byelorussia and Ukraine have been evacuated and another 110,000 await relocation. Yet 5 million people live in the overall affected area. Contamination is continuing to spread through wind, air, water, soil, and food chains. Unstable isotopes of plutonium, americium, cesium, and strontium pose a distinct threat to the population. (Source: Dr. Yevgeny

F. Konoplya, Director of the Radiobiology Institute of the Byelorussian Academy of Sciences.) Clearly, one person or country's exhalation becomes another's inhalation downwind!

• *Define* wind and *describe* how wind is measured, how wind direction is determined, and how winds are named.

5. Define wind. How is it measured? How is its direction determined?

Wind is the horizontal motion of air relative to Earth's surface. It is produced by differences in air pressure from one location to another and is influenced by several variables. The two principal variables are speed and direction. Wind speed is measured with an anemometer, and may be expressed in kilometers per hour (kmph), miles per hour (mph), meters per second (m/sec), or knots. (A knot is a nautical mile per hour, covering 1 minute of Earth's arc in an hour, equivalent to 1.85 kmph or 1.15 mph.) Wind direction is determined with a wind vane; the standard measurement is taken 10 m (33 ft) above the ground to avoid, as much as possible, local effects of topography upon wind direction.

6. Distinguish among primary, secondary, and tertiary classifications of global atmospheric circulation.

Primary (general) circulation, secondary circulation of migratory high-pressure and low-pressure systems, and tertiary circulation that includes local winds and temporal weather patterns are three classes of global winds. Winds that move principally north or south along meridians are known as meridianal flow or meridianal circulations. Winds moving east or west along parallels of latitude are called zonal flows, or zonal circulations.

7. How was the image in Figure 4.6a produced? What does it demonstrate about winds in the Pacific?

See Figure 4.6a and read caption.

• *Explain* the four driving forces within the atmosphere—gravity, pressure gradient force, Coriolis force, and friction force—and *describe* the primary high- and low-pressure areas and principal winds.

8. What does an isobaric map of surface air pressure portray? Contrast pressures over North America for January and July.

An isobaric map portrays the weight of the atmosphere in specific locations. By delineating high and low pressure air masses, geographers are able to determine the movement of air caused by the pressure gradient force, and determine the stability of an air mass, predicting patterns of precipitation and aridity. Figure 4.11 illustrates the extremes in pressure and how they relate to predictable weather patterns.

Pressures over North America during January reflect the heat capacity of water. Low pressures, such as the Aleutian Low and the Icelandic Low are found over the Pacific and Atlantic Oceans, respectively. During January, high pressures dominate the North American land mass due to extreme low temperatures, correlate with Figure 3.24. In July these pressures switch locations, due to the low specific heat of land, the North American landmass heats rapidly, causing low pressures to be located over the North American continent and high pressures over the oceans (Figure 3.26). The stable, high pressure air limits much precipitation on the West coast, while warm, low pressure air dominates the Eastern Seaboard, causing greater evaporation rates, higher humidity, and summer showers.

9. Describe the effect of the Coriolis force. Explain how does it apparently deflects atmospheric and oceanic circulations.

The Coriolis force applies to objects moving across Earth's surface that appear to deflect from a straight path. Because the physicist Sir Isaac Newton stated that if something is accelerating over a space, a force is in operation, the label *force* is appropriate here. The deflection produced by the Coriolis force is caused by the fact that Earth's rotational speed varies with latitude, decreasing from 1675 kmph (1041 mph) at the equator to 0 at the poles. The Coriolis effect increases as the speed of moving objects increases; thus, the faster the movement, the greater the apparent deflection. And because Earth rotates eastward, objects that move in an absolute straight line over a distance for some time (such as winds and ocean currents) appear to curve to the right in the Northern Hemisphere and to the left in the Southern Hemisphere.

Do we call Coriolis a *force* or an *effect*? The proper way to refer to the Coriolis force has plagued physical geography texts for years, yet in other academic areas such ambivalence and confusion do not seem to appear. I chose to add the reference to Newton's definition of a force to give you a way of explaining the usage in *Elemental Geosystems*. Think of a deflective force that causes these apparent effects. For several basic examples of the typical treatment of Coriolis in other works consult the following among many. I briefly quote these sources to give you an idea of usage.

"Coriolis force," *Encyclopedia Britannica*, Micropædia Vol. 3, Chicago (1989 ed.): 632. "...in classical mechanics, an inertial force..."

Barry Roger B. and Richard J. Chorley, *Atmosphere, Weather, and Climate*, 5th ed., London: Methuen, 1987, 116–118. "The Coriolis force arises from the fact that the movement of masses over the earth's surface is usually referred to as a moving coordinate system...."

Petterssen, Sverre, *Introduction to Meteorology*, 3rd ed., New York: McGraw-Hill Book Company, 1969, 153–155. "The deviating force...."

Neiburger, Morris and James G. Edinger, William D. Bonner, *Understanding Our Atmospheric Environment*, San Francisco: W.H. Freeman, 1973, 99–104. "The effect of the rotation of the earth on any moving object is to make it appear as though a force is acting on the object. This apparent force is called the Coriolis force...." (A quantified explanation and description of angular velocity principles is presented).

10. What are geostrophic winds, and where are they encountered in the atmosphere?

Figure 4.8b, illustrates the combined effect of the pressure gradient force and the Coriolis force on the atmosphere, producing geostrophic winds. Geostrophic winds are characteristic of upper tropospheric circulation. The air does not flow directly from high to low, but around the pressure areas instead, remaining parallel to the isobars and producing the characteristic pattern shown on the upper-air weather chart.

11. Describe the horizontal and vertical air motions in a high-pressure anticyclone and in a low-pressure cyclone.

The pressure gradient force acting alone is shown in Figure 4.8a from two perspectives. As air descends in the high-pressure area, a field of subsiding, or sinking, air develops. Air moves out of the high-pressure area in a flow described as diverging. High-pressure areas feature descending, diverging air flows. On the other hand, air moving into a low-pressure area does so with a converging flow. Thus, low-pressure areas feature converging, ascending air flows.

12. Construct a simple diagram of Earth's general circulation, including the four principal pressure belts or zones and the three principal wind systems.

For the student to complete (Figures 4.13 and Table 4.1).

13. How is the intertropical convergence zone (ITCZ) related to the equatorial low-pressure trough? How might it appear on a satellite image?

The equatorial low-pressure trough is an elongated, narrow band of low pressure that nearly girdles Earth, following an undulating linear axis. Constant high Sun altitude and consistent daylength make large amounts of energy available at this region of Earth's surface throughout the year. The warming creates lighter, less-dense, ascending air, with winds converging all along the extent of the trough. The combination of heating and convergence forces air aloft and forms the intertropical convergence zone (ITCZ). This converging air is extremely moist and full of latent heat energy. Vertical cloud columns frequently reach the tropopause, and precipitation is heavy throughout this zone. On a satellite image it is identified by bands of clouds along the equator.

14. Characterize the belt of subtropical high pressure on Earth: Name the specific cells. Describe the generation of westerlies and trade winds. Discuss sailing conditions.

Subtropical high pressure consists of hot, dry air resulting from air diverging from the Hadley cell deflected to the poles by the Coriolis force. This air is associated with cloudless, desert regions, such as the Sahara and the Arabian deserts. Examples of subtropical high pressure cells are: the Azores high, which dominates the west coast of Africa in the Atlantic Ocean; the Bermuda High, located in the western Atlantic; and the Pacific High, which dominates the Pacific Ocean. The divergence of this dry air creates the ocean gyres, spinning in a clockwise direction in the North Hemisphere due to the Coriolis force.

The ocean currents stimulated by the divergence of high pressure create two regions of predictable currents, the trade winds, which migrate in an easterly direction, creating currents such as the Canary current and the Equatorial Counter Current between Europe and North America. And the westerlies, which move in a westerly direction causing such currents as the Gulf Stream in the Atlantic and the Kurioshio, Japan Current, in the Pacific. The combination of trade winds, or easterlies, and westerlies, enabled European sailors to travel to North America and return to Europe using the currents created by the subtropical highs. If ships were not geographically accurate, they may have gotten caught in the center of the subtropical highs, which are characterized by little wind. These areas are called the *horse latitudes*, where ships often killed livestock in order to ration water and food, fearing they would not locate the trade winds.

15. What is the relation among the Aleutian low, the Icelandic low, and migratory low-pressure cyclonic storms in North America? In Europe?

The Aleutian low and Icelandic low are migratory pressure cells. During the summer, these low pressure cells are found in high latitudes over the Pacific and Atlantic Oceans, bring precipitation to Ireland and the Pacific Northwest, such as Seattle and Alaska. In January, as the continent of North America becomes dominated by high pressure, these subpolar lows migrate to lower latitudes (due to their relative temperature), causing cyclonic storms of the West Coast of North America and Europe, such as California and Spain. The contrasts that these areas experience, variations between high pressure in the winter and low pressures in the summer, cause frontal interaction and predictable patterns of precipitation.

• *Describe* upper-air circulation and its support role for surface systems and *define* the jet streams.

16. What is the relation between wind speed and the spacing of isobars?

Isobars allow us to see the pressure gradient that exists in specific locations in the atmosphere. When the isobars are close together, this shows a steep pressure gradient, or difference, causing greater wind speeds. Similar to a topographic map, where closely clustered lines of elevation show a steep slope, we expect to find strong winds in areas of closely clustered isobars. If isobars are spaced far apart, this reflects a gentle pressure gradient resulting in a slow wind speed, or a mild breeze. See Figure 4.7.

17. How are undulations in the upper-air circulation (ridges and troughs) related to surface pressure systems? To divergence aloft and surface lows? To convergence aloft and surface highs?

See Figure 4.10 and the text. As on surface maps, closer spacing of the isobars indicates faster winds; wider spacing indicates slower winds. On this isobaric pressure surface, altitude variations from the reference datum are called ridges for high pressure, and troughs for low pressure. The patterns of ridges and troughs in the upper-wind flow are important in sustaining surface cyclonic (low) and anticyclonic (high) circulation. Frequently surface pressure systems are generated by the upper-air wind flow. Near ridges in the isobaric surface, winds slow and converge. Conversely, near the area of maximum wind speeds in the isobaric surface, winds accelerate and diverge. This divergence in the upper-air flow is important to cyclonic circulation at the surface because it creates an outflow of air aloft that stimulates an inflow of air into the low-pressure cyclone.

18. Relate the jet-stream phenomenon to general upper-air circulation. How is the presence of this circulation related to airline schedules from New York to San Francisco and the return trip to New York?

Within the westerly flow of geostrophic winds are great waving undulations called Rossby waves (Figure 4.16). These Rossby waves develop along the flow axis of a jet stream. The most prominent movement in these upper-level westerly wind flows is the jet stream, an irregular, concentrated band of wind occurring at several different locations. Airline schedules reflect the presence of these upper-level winds, for they allot shorter flight times from west to east and longer flight times from east to west (Figure 4.17, p. 123). See News Report 4.2.

• Overview several multiyear oscillations of air temperature, air pressure, and circulation in the Arctic, Atlantic, and Pacific Oceans.

19. What phases are identified for the AO and NAO indices? What winter weather conditions generally affect the eastern United States during each phase?

A north–south fluctuation of atmospheric variability marks the North Atlantic Oscillation (NAO) as pressure differences between the Icelandic low and the Azores high in the Atlantic alternate in strength. The Atlantic Oscillation (AO) is the variable fluctuation between middle- and high-latitude air mass conditions over the Northern Hemisphere. The AO is associated with the NAO especially in winter, and the two phases of the AO Index correlate to the two-phased NAO Index.

20. What is the apparent relation between the PDO and the strength of El Nino events? Between PDO phases and the intensity of drought in the southwestern United States?

Across the Pacific Ocean, the Pacific Decadal Oscillation (PDO) is longer lived, at 20 to 30 years in duration, than the 2-to 12-year variation in the ENSO. The PDO term came into use in 996. Involved are two regions of sea-surface temperatures and related air pressure: the northern and tropical western Pacific (region #1), and the area of the eastern tropical Pacific, along the West Coast (region #2).

Between 1947 and 1977, higher-than-normal temperatures dominated region #1, and lower temperatures were in region #2; this is the PDO negative phase (or cool phase). A switch to a positive phase (or warm phase) in the PDO ran from 1977 to the 1990s, when lower temperatures were in region #1 and higher-than-normal temperatures dominated region #2, coinciding with a time of more intense ENSO events. In 1999, a switch to a negative phase began. Unfortunately for the U.S. Southwest, this PDO negative phase can mean a decade or more of drier conditions for the already drought-plagued region.

Causes of the PDO and its cyclic variability over time are unknown. Scientists monitor conditions in the Pacific and look for patterns. A better understanding of the PDO will help scientists predict ENSO events as well as regional drought cycles.

• *Explain* several types of local winds: land-sea breezes, mountain-valley breezes, katabatic winds, and the regional monsoons.

21. People living along coastlines generally experience variations in winds from day to night. Explain the factors that produce these changing wind patterns.

The differential heating characteristics of land and water surfaces create these winds. During the day, land heats faster and becomes warmer than does the water offshore. Because warm air is less dense, it rises and triggers an onshore flow of cooler marine air, which is usually stronger in the afternoon. At night, inland areas cool radiatively faster than do offshore waters. As a result, the cooler air over the land subsides and flows offshore over the warmer water, where the air is lifted. This night pattern reverses the process that developed during the day (Figure 4.18).

22. The arrangement of mountains and nearby valleys produces local wind patterns. Explain the day and night winds that might develop.

Mountain air cools rapidly at night, and valley air gains heat rapidly during the day (Figure 4.19). Thus, warm air rises upslope during the day, particularly in the afternoon; at night, cooler air subsides downslope into the valleys.

23. Describe the seasonal pressure patterns that produce the Asian monsoonal wind and precipitation patterns. Contrast January and July conditions.

The monsoons of southern and eastern Asia are driven by the location and size of the Asian landmass, and its proximity to the Indian Ocean and the seasonally shifting ITCZ (Figure 4.20). Also important to the generation of monsoonal flows are wind and pressure patterns in the upper-air circulation. The extreme temperature range from summer to winter over the Asian landmass is due to its continentality, reflecting its isolation from the modifying effects of the

ocean. Resultant cold, dry winds blow from the Asian interior over the Himalayas, downslope, and across India, producing average temperatures of between 15° and 20°C (60° and 68°F) at lower elevations. These dry winds desiccate the landscape.

During the June–September wet period, the Sun shifts northward to the Tropic of Cancer, near the mouths of the Indus and Ganges rivers. The Asian continental interior develops a thermal low pressure, associated with higher average temperatures, and the intertropical convergence zone shifts northward over southern Asia. Meanwhile, the Indian Ocean, with surface temperatures of 30°C (86°F), is under the influence of subtropical high pressure. These conditions produce the wet monsoon of India, where world-record rainfalls have occurred: both the second-highest average rainfall (1143 cm or 450 in.) and the highest single-year rainfall total (2647 cm or 1042 in.), were measured at Cherrapungi, India.

• *Describe* the basic pattern of Earth's major surface and deep ocean currents.

24. What is the relationship between global atmospheric circulation and ocean currents? Relate oceanic gyres to patterns of subtropical high pressure.

Earth's atmospheric and oceanic circulations are interrelated. The driving force for ocean currents is the frictional drag of the winds. Also important in shaping these currents is the interplay of the Coriolis force, density differences associated with temperature and salinity, the configuration of the continents and ocean floor, and astronomical forces (the tides). Ocean currents are driven by atmospheric circulation around subtropical high-pressure cells in both hemispheres. These circulation systems are known as gyres and generally appear offset toward the western side of each ocean basin.

25. Define the western intensification. How is it related to the Gulf Stream and Kuroshio currents?

Gyres and their western margins feature slightly stronger currents than do the eastern portions. Along the full extent of areas adjoining the equator, trade winds drive the oceans westward in a concentrated channel. These currents are kept near the equator by the weaker Coriolis force influence. As the surface current approaches the western margins of the oceans, the water actually piles up an average of 15 cm (6 in.). From this western edge, ocean water then spills northward and southward in strong currents, flowing in tight channels along the western edges of the ocean basins (eastern shorelines of continents). This is the process known as western intensification.

In the Northern Hemisphere, the Gulf Stream and the Kuroshio move forcefully northward as a result of western intensification, with their speed and depth increasing with the constriction of the area they occupy. The warm, deep-blue water of the ribbonlike Gulf Stream (Figure 3.20) usually is 50–80 km wide and 1.5-2.0 km deep (30–50 mi wide and 0.9–1.2 mi deep), moving at 3–10 kmph (1.8–6.2 mph). In 24 hours, ocean water can move 70–240 km (40–150 mi) in the Gulf Stream, although a complete circuit around an entire gyre may take a year.

26. Where on Earth are upwelling currents experienced? What is the nature of these currents?

Where surface water is swept away from a coast, either by surface divergence (induced by the Coriolis force) or by offshore winds, an upwelling current occurs. This cool water generally is nutrient-rich and rises from great depth to replace the vacating water. Such cold upwelling currents occur off the Pacific coasts of North and South America and the subtropical and midlatitude west coast of Africa.

27. What is meant by deep-ocean circulation? At what rates do these currents flow? How might this circulation be related to the Gulf Stream in the western Atlantic ocean?

Important mixing currents along the ocean floor are generated from such downwelling zones and travel the full extent of the ocean basins, carrying heat energy and salinity.

Imagine a continuous channel of water beginning with cold water downwelling in the North Atlantic, flowing deep and strong to upwellings in the Indian Ocean and North Pacific. Here it warms and then is carried in surface currents back to the North Atlantic. A complete circuit may require 1000 years from downwelling in the Labrador Sea off Greenland to its reemergence in the southern Indian Ocean and return. Even deeper Antarctic bottom water flows northward in the Atlantic Basin beneath these currents.

27. Relative to Question 27, what relation do these deep currents have to global warming and possible climate change?

These current systems appear to play a profound role in global climate; in turn, global warming has the potential to disrupt the downwelling in the North Atlantic and the thermohaline circulation. If this sinking of surface waters slowed or stopped, the ocean's ability to redistribute absorbed heat energy would be disrupted and surface accumulation of heat energy would add to the warming already under way. There is a freshening of ocean surface waters in both polar regions, contrasted with large increases in salinity in the surface waters at lower latitudes. Increased rates of glacier, sea-ice, and ice-sheet melting produce these fresh lower-density surface waters that ride on top of the denser saline water. The concern is that such changes in ocean temperature and salinity could dampen the rate of the North Atlantic Deep Water downwelling in the North Atlantic. One study published in November 2005 reported to have detected a slowing of 30% from the long-term average. Such slowing could send Europe into a lower-than-normal temperature regime.

Overhead Transparencies

	Chapter 4	Atmospheric and Oceanic Circulation
67.	4.1 a – e	Volcanic eruption effects spread worldwide by winds
68.	4.2 a, b	Developing the barometer
69.	4.3	Air pressure readings and conversions
70.	4.5	A wind compass
71.	4.6 a, b	*SEASAT* surface winds; Earth image
72.	4.7 a, b	Pressure gradient determines wind speed (top); surface weather map (bottom)
73.	4.8 a, b, c	Three physical forces that produce winds
74.	4.9 a, b, c	Coriolis force
75.	4.10	Isobaric chart (500-mb) for an April day
76.	4.11 a, b	Global barometric pressures for January (top); and July (bottom)
77.	4.13 a, b	General atmospheric circulation model (top); atmospheric profile (bottom)
78.	4.14	Subtropical high pressure system in the Atlantic
79.	4.16 a, b, c	Development Of Rossby waves in upper atmosphere
80.	4.17a, b; N.R. 4.1.1	Location of 2 jet streams; polar jet stream (top); map of MANTRA's drift path (bottom)
81.	4.18	Land-sea breezes; pressure surfaces
82.	4.19	Mountain-valley breezes
83.	4.20 a, b, c	The Asian monsoon, Dec-Jan (a), July-Aug (b), Nagpur climograph (c)
84.	4.21	Ocean currents
85.	4.22	Deep ocean circulation
86.	F.S. 4.1.1, and F.S. 4.1.2 a, b	Wind-generated power 1980–2004 (top); wind-power installations (bottom)
87.	N.R. 4.3.1	Pacific ocean currents transports human artifacts

PART TWO:
Water, Weather, and Climate Systems

Overview—Part Two

Part Two presents aspects of hydrology, meteorology and weather, oceanography, and climate, in a flowing sequence in Chapters 5 through 7. Earth is the water planet among planets in the Solar System. We begin with water itself—its origin, location, properties, and dynamic circulation in the hydrologic cycle. The global oceans and seas are identified as the greatest repository of water on Earth. The dynamics of daily weather phenomena follow: the effects of moisture and energy in the atmosphere, the interpretation of cloud forms, conditions of stability or instability, the interaction of air masses, and the occurrence of violent weather. We examine water resources—the specifics of the hydrologic cycle and how water circulates over Earth. We look at the water-balance concept as a useful way to understand water-resource relationships, whether global, regional, or local. Important water resources include rivers, lakes, groundwater, and oceans. As an ultimate output of these systems, we see the spatial implications over time of the energy-atmosphere and water-weather systems that generate Earth's climate patterns. And we analyze the human impact on the climate system. This "pale blue dot" in space is indeed the water planet.

Atmospheric Water and Weather

Water is the essential medium of our daily lives and a principal compound in nature. It covers 71% of Earth (by area) and in the solar system occurs in such significant quantities only on our planet. It constitutes nearly 70% of our bodies by weight and is the major component in plants and animals. The water we use must be adequate in quantity as well as quality for its many tasks—everything from personal hygiene to vast national water projects. Indeed water occupies the place between land and sky, mediating energy and shaping both the lithosphere and atmosphere, as stated in the text.

This chapter examines the dynamics of water and atmospheric moisture in particular. Discussion of the water, weather, and climate system, starts with the beginning of Earth's water. The text then follows the accumulation of water, and describes the unique properties of water.

Our study of weather begins with a discussion of atmospheric stability and instability. The clouds that form as air becomes water-saturated are more than whimsical, beautiful configurations; they are important indicators of overall atmospheric conditions.

We follow huge air masses across North America, observe powerful lifting mechanisms in the atmosphere, examine cyclonic systems, and conclude with a portrait of the violent and dramatic weather that occurs in the atmosphere. Temperature, air pressure, relative humidity, wind speed and direction, daylength, and Sun angle are important measurable elements that contribute to the weather. We tune to local stations for the day's weather report from the National Weather Service (in the United States) or the Meteorological Service of Canada to see the current satellite images and to hear tomorrow's forecast.

Outline Headings and Key Terms

The first-, second-, and third-order headings that divide Chapter 5 serve as an outline for your notes and studies. The key terms and concepts that appear boldface in the text are listed here under their appropriate heading in **bold italics**. All these highlighted terms appear in the text glossary. Note the check-off box (❑) so you can mark your progress as you master each concept. Your students have this same outline in their *Student Study Guide*. The ✪ icon indicates that there is an accompanying animation or other resource on the CD.

The outline headings for Chapter 5:

Water on Earth: Location and Properties
✪ **Earth's Water and the Hydrologic Cycle**
    ❑ *outgassing*
Quantity Equilibrium
Distribution of Earth's Water
Unique Properties of Water
✪ **Water Phase Changes**
Heat Properties
    ❑ *phase change*
    ❑ *sublimation*
Ice, the Solid Phase
Water, the Liquid Phase
    ❑ *latent heat*
Water Vapor, the Gas Phase
    ❑ *latent heat of vaporization*
    ❑ *latent heat of condensation*
Heat Properties of Water in Nature
    ❑ *latent heat of evaporation*
Humidity
    ❑ *humidity*
Relative Humidity
    ❑ *relative humidity*
    ❑ *saturated*
    ❑ *dew-point temperature*
Expressions of Relative Humidity
Vapor Pressure
    ❑ *vapor pressure*
Specific Humidity
    ❑ *specific humidity*
Instruments for Measurement
Atmospheric Stability
✪ **Atmospheric Stability**
    ❑ *stability*
Adiabatic Processes
    ❑ *normal lapse rate*
    ❑ *environmental lapse rate*
    ❑ *adiabatic*
Dry Adiabatic Rate (DAR)
    ❑ *dry adiabatic rate*
Moist Adiabatic Rate (MAR)

    ❑ *moist adiabatic rate*
Stable and Unstable Atmospheric Conditions
Clouds and Fog
    ❑ *cloud*
    ❑ *moisture droplets*
    ❑ *condensation nuclei*
Cloud Types and Identification
    ❑ *stratus*
    ❑ *nimbostratus*
    ❑ *cumulus*
    ❑ *stratocumulus*
    ❑ *altocumulus*
    ❑ *cirrus*
    ❑ *cumulonimbus*
Fog
Advection Fog
    ❑ *advection fog*
    ❑ *evaporation fog*
    ❑ *upslope fog*
    ❑ *valley fog*
Radiation Fog
    ❑ *radiation fog*
Air Masses
    ❑ *air mass*
Air Mass Modification
Atmospheric Lifting Mechanisms
✪ **Convectional Heating and Tornado in Florida**
✪ **Cold and warm fronts**
    ❑ *convergent lifting*
Convectional Lifting
    ❑ *convectional lifting*
Orographic Lifting
    ❑ *orographic lifting*
    ❑ *rain shadow*
Frontal Lifting (Warm and Cold Fronts)
    ❑ *cold front*
    ❑ *warm front*
Cold Front
Warm Front
Midlatitude Cyclonic Systems
✪ **Midlatitude Cyclones**
    ❑ *weather*
    ❑ *meteorology*
    ❑ *midlatitude cyclone*
    ❑ *wave cyclone*
Life Cycle of a Midlatitude Cyclone
Storm Tracks
Open Stage
Occluded Stage
    ❑ *occluded front*
Violent Weather
✪ **Tornado Wind Patterns**
✪ **Hurricane Wind Patterns**
✪ **Hurricane Isabel 9/6-9/19/2004**
✪ **2004 Florida Hurricanes Image and Movies:**
Hurricane Charley 8/9–8/15/04 (satellite montage)

The URLs related to this chapter of *Elemental Geosystems* can be found at
http://www.prenticehall.com/christopherson

Key Learning Concepts

After reading the chapter and using this study guide, the student should be able to:

• *Describe* the origin of Earth's waters, the present distribution, and water's unique heat properties.

• *Define* humidity, relative humidity, and dew-point temperature and *illustrate* three atmospheric conditions—unstable, conditionally unstable, and stable.

• *Explain* cloud formation and *identify* major cloud classes and types, including fog.

• *Describe* air masses that affect North America and *identify* four types of atmospheric lifting mechanisms.

• *List* the measurable elements of weather and *describe* the life cycle of a midlatitude cyclone.

• *Analyze* various types of violent weather— thunderstorms, tornadoes, and tropical cyclones.

Annotated Chapter Review Questions

• *Describe* the origin of Earth's waters, the present distribution, and water's unique heat properties.

1. Approximately where and when did Earth's water reach its present volume?

All water on Earth was formed within the planet, reaching Earth's surface in an on-going process called outgassing, by which water and water vapor emerge from layers deep within and below the crust, as much as 25 km (15.5 mi) or more below Earth's surface (Figure 5.1). Various geophysical factors explain the timing of this outgassing over the past 4 billion years. In the early atmosphere, massive quantities of outgassed water vapor condensed and fell in torrents, only to vaporize again because of high temperature at Earth's surface. For water to remain on Earth's surface, land temperatures had to drop below the boiling point of 100°C (212°F), something that occurred about 3.8 billion years ago.

2. If the quantity of water on Earth has been quite constant in volume for at least 2 billion years, how can sea level have fluctuated? Explain.

Today, water is the most common compound on Earth, having achieved the present volume of 1.36 billion km^3 (326 million mi^3) approximately 2 billion years ago. This quantity has remained relatively constant, even though water is continuously being lost from the system, escaping to space or breaking down and forming new compounds with other elements. Lost water is replaced by pristine water not previously at the surface, water that emerges from within Earth.

The net result of these inputs and outputs to water quantity is a steady-state equilibrium in Earth's hydrosphere. Despite this overall net balance in quantity, worldwide changes in sea level do occur and are called eustasy, which is specifically related to changes in volume of water and not movement of land. These changes are explained by glacio-eustatic factors (see Chapter 14). Glacio-eustatic factors are based on the amount of water stored on Earth as ice. Over the past 100 years, mean sea level has steadily risen and is still rising worldwide at this time. Apparent changes in sea level also are related to actual physical changes in landmasses called isostatic change, such as continental uplift or subsidence (isostasy is discussed in Chapter 8).

3. Describe the location of Earth's water, both oceanic and fresh. What is the largest repository of freshwater at this time? In what ways is this distribution significant to modern society?

See Figure 5.3. The greatest single repository of surface freshwater is ice. Ice sheets and glaciers account for 77.78% of all freshwater on Earth. Add to this the subsurface groundwater, and that accounts for 99.36% of all freshwater. The remaining freshwater, although very familiar to us, and present in seemingly huge amounts in lakes, rivers, and streams, actually represents but a small quantity, less than 1%.

The southern portions of the Indian, Atlantic, and Pacific are sometimes called the Southern Ocean. See the Institute of Oceanographic Sciences, Deacon Laboratory. The FRAM (Fine Resolution Antarctic Model) Atlas of the Southern Ocean. Wormley, Surrey: Natural Environment Research Council, 1991, for detailed description of currents and extent of the Southern Ocean.

4. Why might you describe Earth as the water planet? Explain.

Earth's waters exist in the atmosphere, on the surface, and in the crust near the surface, in liquid, solid, and gaseous forms. Water occurs as fresh and saline, and exhibits important heat properties as well as an extraordinary role as a solvent. Among the planets in the Solar System, only Earth possesses water in such quantity.

5. Describe the three states of matter as they apply to ice, water, and water vapor.

Earth's distance from the Sun places it within a most remarkable temperate zone compared with the other planets. This temperate location allows all states of water to occur naturally on Earth: ice, in the solid phase; water, in the liquid phase; and water vapor, in the gas phase. See Figures 5.4 and 5.6.

6. What happens to the physical structure of water as it cools below 4°C (39°F)? What are some of the visible indications of these physical changes?

As water cools, it behaves like most compounds and contracts in volume, reaching its greatest density at 4°C (39°F). But below that temperature, water behaves very differently from most compounds, and begins to expand as more hydrogen bonds form among the slower-moving molecules, creating the hexagonal structures shown in Figure 5.4. This expansion continues to a temperature of –29°C (–20°F), with up to a 9% increase in volume possible. As shown in Figure 5.5a, the rigid internal structure of ice dictates the six-sided appearance of all ice crystals, which can loosely combine to form snowflakes. The expansion in volume that accompanies the freezing process results in a decrease in density. Specifically, ice has 0.91 times the density of water, and so it floats. The freezing action of ice as an important physical weathering process is discussed in Chapters 10 and 14.

7. What is latent heat? How is it involved in the phase changes of water?

For water to change from one state to another, heat energy must be added to it or released from it. The amount of heat energy must be sufficient to affect the hydrogen bonds between the molecules. For ice to melt, heat energy must increase the motion of the water molecules to break some of the hydrogen bonds. Despite the fact that there is no change in sensible temperature between ice at 0°C and water at 0°C, 80 calories are required for the phase change of 1 gram of ice to 1 gram of water. This heat, called *latent heat*, is hidden within the water and is liberated whenever a gram of water freezes. To accomplish the phase change of liquid to vapor at boiling, under normal sea-level pressure, 540 calories must be added to 1 gram of boiling water to achieve a phase change to water vapor. Those calories are the latent heat of vaporization. See Figure 5.6.

8. Take 1 g of water at 0°C and follow it through to 1 g of water vapor at 100°C, describing what happens along the way. What amounts of energy are involved in the changes that take place?

See Figure 5.6. If we take a gram of water at O°C, it will require 100 calories to raise the temperature of water to 100°C, one calorie for each degree of change in temperature. At 100°C, water is at its boiling point. To change water's phase from liquid to vapor, will require another 540 calories. This energy is called the latent heat of vaporization. The total amount of energy absorbed by water in this process was 640 calories, 100 cal + 540 cal = 640 cal. To take a gram of ice at 0°C and turn it into vapor at 100°C, would require 80 additional calories, this energy absorbed by the water is called the latent heat of melting, or the latent heat of fusion. Thus, to transform ice at 0°C to vapor at 100°C would require a total of 720 calories, 80 calories to change from ice at 0°C to water at 0°C, plus 100 calories to heat water from 0°C to 100°C, and finally the additional 540 calories required to heat water at 100°C to vapor at 100°C; 80 cal + 100 cal + 540 cal = 720 cal.

• Define humidity, relative humidity, and dew-point temperature and illustrate three atmospheric conditions—unstable, conditionally unstable, and stable.

9. What is humidity? How is it related to the energy present in the atmosphere? To our personal comfort and how we perceive apparent temperatures?

Humidity is the vapor content of air. The vapor content of air changes according to the temperature of air and the temperature of water vapor. Warm air has a greater capacity to hold water than does cold air. Warm air stores a greater amount of energy in the atmosphere as the latent heat of evaporation, than cold air. Humidity affects the apparent temperature that we experience. For example, air with very little humidity, while it may be hot, is much more comfortable for humans. The warmth of the air enables the atmosphere to hold more moisture, and as we perspire, our bodies are cooled by this heat transfer. If we live in areas that are hot and humid, this means that the air is close to being saturated and may not be able to hold more moisture. Our perspiration may not be absorbed by the atmosphere, making the apparent heat seem much higher than actual air temperature.

We humans attempt to adjust the humidity levels in our homes through humidifiers, which add vapor to the air, increasing apparent temperatures in the winter; and through the use of air conditioners, we can cool the air, extracting the vapor from the atmosphere, reducing the apparent temperatures.

10. Define relative humidity. What does the concept represent? What is meant by the terms *saturation* and *dew-point temperature*?

The water vapor content of air is termed humidity. The capacity of air to hold water vapor is primarily a function of temperature: warmer air has a greater capacity for water vapor, whereas cooler air has a lesser capacity. Relative humidity is not a direct measurement of water vapor; rather, it is expressed as a percentage of the amount of water vapor that is actually in the air (content), compared with the maximum water vapor the air could hold at a given temperature (capacity). Air is said to be saturated, or full, if it is holding all the water vapor that it can hold at a given temperature; under such conditions, the net transfer of water molecules between surface and air achieves a saturation equilibrium. Saturation indicates that any

further addition of water vapor (change in content) or any decrease in temperature (change in capacity) will result in active condensation.

The temperature at which a given mass of air becomes saturated is termed the dew-point temperature. In other words, air is saturated when the dew-point temperature and the air temperature are the same.

11. Using several different measures of humidity in the air given in the chapter, derive relative humidity values (vapor pressure/saturation vapor pressure; specific humidity/maximum specific humidity).

As the Figure 5.11 graph shows, air at 20°C (68°F) has a saturation vapor pressure of 24 mb; that is, the air is saturated if the water vapor portion of the air pressure is at 24 mb. Thus, if the water vapor content actually present is exerting a vapor pressure of only 12 mb in 20°C air, the relative humidity is only 50% (12 mb ÷ 24 mb = 0.50 x 100 = 50%). The graph illustrates that, for every temperature increase of 10C° (18F°), the vapor pressure capacity of air nearly doubles.

The maximum mass of water vapor that a kilogram of air can hold at any specified temperature is termed the maximum specific humidity and is plotted in Figure 5.12. This graph shows that a kilogram of air could hold a maximum specific humidity of 47 g of water vapor at 40°C (104°F), 15 g at 20°C (68°F), and 4 g at 0°C (32°F). Therefore, if a kilogram of air at 40°C has a specific humidity of 12 g, its relative humidity is 25.5% (12 g + 47 g = 0.255 x 100 = 25.5%). Specific humidity is useful in describing the moisture content of large air masses that are interacting in a weather system.

12. How do the two instruments described in this chapter measure relative humidity?

The hair hygrometer uses the principle that a human hair changes as much as 4% in length between 0 and 100% relative humidity. The instrument connects a standardized bundle of human hair through a mechanism to a gauge and a graph to indicate relative humidity. Another instrument used to measure relative humidity is a sling psychrometer. Figure 5.13b shows this device, which has two thermometers mounted side-by-side on a metal holder. One is called the dry-bulb thermometer; it simply records the ambient air temperature. The other thermometer is called the wet-bulb thermo-meter; it is set lower in the holder and has a cloth wick over the bulb, which is moistened with distilled water. The psychrometer is spun by its handle for a minute or two. The readings on the two thermometers can then be checked on a psychrometric table to determine relative humidity. The greater the difference between the two thermometers, the lower the relative humidity.

13. Differentiate between stability and instability relative to a parcel of air rising vertically in the atmosphere.

Stability refers to the tendency of a parcel of air to either remain as it is or change its initial position by lifting or falling. An air parcel is termed stable when it resists displacement upward or, if disturbed, it tends to return to its starting place. On the other hand, an air parcel is considered unstable when it continues to rise until it reaches an altitude where the surrounding air has a density similar to its own. Determining the degree of stability or instability involves measuring simple temperature relationships between the air parcel and the surrounding air.

14. What are the forces acting on a vertically moving parcel of air? How are they affected by the density of the air parcel?

There are two opposing forces which act on a vertically moving parcel of air. These forces are an upward buoyant force and a downward gravitational force, Figure 5.14. Temperature and density characteristics of the air mass determine which force will be more influential. If an air mass is higher in temperature than the surrounding atmosphere or if an air mass is less dense than the surrounding air, it will continue to rise vertically. Similar to any gas, as it rises in the air, the air mass begins to expand due to decreasing pressure in higher altitudes of the atmosphere. An unstable air mass will continue to rise until the surrounding air has similar density and temperature characteristics. An air mass will become dominated by the downward pull of gravity as it cools in the atmosphere. The temperature and density of the air mass may be lower than the surrounding air. This will cause the air mass to descend. As the parcel falls, external pressure in the atmosphere increases causing the air mass to compress.

15. How do the adiabatic rates of heating or cooling in a vertically displaced air parcel differ from the normal lapse rate and environmental lapse rate?

The normal lapse rate, as introduced in Chapter 2, is the average decrease in temperature with increasing altitude, a value of 6.4°C per 1000 m (3.5°F per 1000 ft). This rate of temperature change is for still, calm air, but it can differ greatly under varying weather conditions, and so the actual lapse rate at a particular place and time is labeled the environmental lapse rate. Within this environment, an ascending (lifting) parcel of air tends to cool by expansion, responding to the reduced pressure at higher altitudes. Descending (subsiding) air tends to heat by compression.

The dry adiabatic rate is the rate at which dry air cools by expansion (if ascending) or heats by compression (if descending). Dry air is less than

saturated, with a relative humidity of less than 100%. The DAR is 10°C per 1000 m (5.5°F per 1000 ft), as illustrated in Figure 5.14. The moist adiabatic rate is the average rate at which air that is moist (saturated) cools by expansion (if ascending) or heats by compression (if descending). The average MAR is 6°C per 1000 m (3.3°F per 1000 ft), or roughly 4°C (2°F) less than the DAR. However, the MAR varies with moisture content and temperature, and can range from 4°C to 10°C per 1000 m (2°F to 6°F per 1000 ft).

In a saturated air parcel, latent heat of condensation is liberated as sensible heat, which reduces the adiabatic rate of cooling. The release of latent heat may vary, which affects the MAR. The MAR is much lower than the DAR in warm air, whereas the two rates are more similar in cold air.

16. What would atmospheric temperature and moisture conditions be on a day when the weather is unstable? When it is stable? Relate in your answer what you would experience if you were outside watching.

On a day when the weather is unstable, the atmosphere is dominated by warm air that may absorb available moisture and reach the level of saturation. These warm air parcels may rise vertically, causing them to cool, condense, and precipitate. You would experience evaporation during the warm period of the day, perhaps beginning the morning with a cloudless sky and seeing clouds form throughout the day as the air warms and the humid parcels of air rise in the atmosphere, cooled by the environmental lapse rate. As the parcels cool in the atmosphere, or after 4 P.M., when the atmosphere begins to cool, precipitation may occur.

On a day when the weather is stable, there is no vertical movement of air. The atmosphere remains cool and dry, due to the lower capacity of cool air to hold moisture. You would experience a cloudless sky, or perhaps a greenhouse effect caused by pollution or clouds, which may cool the lower atmosphere.

• *Explain* cloud formation and *identify* major cloud classes and types, including fog.

17. Specifically, what is a cloud? Describe the droplets that form a cloud.

Clouds are not initially composed of raindrops. Instead, they are made up of a multitude of moisture droplets, each individually invisible to the human eye without magnification. An average raindrop, at 2000 μm diameter (0.2 cm or 0.078 in.), is made up of a million or more moisture droplets, each approximately 20 μm diameter (0.002 cm or 0.0008 in.).

18. Explain the condensation process: what are the two essential requirements?

Under unstable conditions, a parcel of air may rise to where it becomes saturated, (i.e., the air cools to the dew-point temperature and 100% relative humidity). Further cooling (lifting) of the air parcel produces active condensation of water vapor to water. This condensation requires microscopic particles called condensation nuclei, which always are present in the atmosphere. In continental air masses, which average 10 billion nuclei per cubic meter, condensation nuclei are derived from ordinary dust, volcanic and forest-fire soot and ash, and particles from combustion.

The collision-coalescence process predominates in clouds that form at above-freezing temperatures, principally in the warm clouds of the tropics. Initially, simple condensation takes place on small nuclei, some of which are larger than others and produce larger water droplets. As those larger droplets fall through a cloud, they combine with smaller droplets, gradually coalescing in size until their weight is beyond the ability of the cloud circulation to hold them aloft. The existence of ice crystals and supercooled water droplets is also a significant mechanism for raindrop production. Figure 5.10 shows that the saturation vapor pressure near an ice surface is lower than that near a water surface; therefore, supercooled water droplets will evaporate more rapidly near ice crystals, which then absorb the vapor. In this ice-crystal process, first proposed in 1928 by Swedish scientist Tor Bergeron, the ice crystals feed on the supercooled cloud droplets, grow in size, and eventually fall as snow or rain. Precipitation in middle and high latitudes begins as ice and snow high in the clouds, then melts and gathers moisture as it falls through the warmer portions of the clouds.

19. What are the basic forms of clouds? Using Figure 5.18, describe how the basic cloud forms vary with altitude.

See Figure 5.18. Clouds usually are classified by altitude and by shape. They come in three basic forms–flat, puffy, and wispy–which are found in four primary classes and ten basic types. Clouds that are developed horizontally and are flat and layered, are called stratiform clouds. Those that are developed vertically and are puffy and globular are termed cumuliform. Wispy clouds usually are quite high, composed of ice crystals, and are labeled cirroform. These three basic forms occur in four altitudinal classes: low, middle, high, and those that are vertically developed occur across all altitudes.

20. Explain how clouds might be used as indicators of the conditions of the atmosphere. Of expected weather?

Clouds can be used to indicate conditions of the atmosphere and expected weather, due to their ability to demonstrate the atmospheric stability, atmospheric temperature, and moisture level or relative humidity. The altitude at which cloud develop are good indicators of atmospheric temperatures. Clouds low in elevation, such as stratus or cumulus clouds, reflect a cool atmosphere, which allow clouds to form at low altitudes in the troposphere. This may lead to condensation and precipitation of a saturated air mass. Cirrus clouds are a good example of this. As they thicken and sink to low elevations, they are often associated with oncoming storms. Vertically developed clouds, such as cumulus or cumulonimbus clouds, demonstrate air masses laden with moisture, and one could expect to see precipitation depending on their height and mass. Other cloud formations illustrate scarcity of moisture in the atmosphere, or a warm atmosphere that limits condensation in the lower atmosphere. Good examples of these clouds could be stratocumulus clouds often associated with clearing weather and wispy cirrus clouds high in the upper troposphere.

21. What type of cloud is fog? List and define the principal types of fog.

A cloud in contact with the ground is commonly referred to as fog. The presence of fog tells us that the air temperature and the dew-point temperature at ground level are nearly identical, producing saturated conditions. Generally, fog is capped by an inversion layer, with as much as $30C°$ $(50F°)$ difference in air temperature between the ground under the fog and the clear, sunny skies above. By international agreement, fog is officially described as a cloud layer on the ground, with visibility restricted to less than 1 km (3300 ft).

As the name implies, *advection fog* forms when air in one place migrates to another place where saturated conditions exist. Another type of fog forms when cold air flows over the warm water of a lake, ocean surface, or even a swimming pool. An *evaporation fog*, or steam fog, may form as the water molecules evaporate from the water surface into the cold overlying air (see Part opening photo). A type of advection fog involving the movement of air forms when moist air is forced to higher elevations along a hill or mountain—an *upslope fog*. This upslope lifting leads to cooling by expansion as the air rises. *Radiation fog* forms when radiative cooling of a surface chills the air layer directly above that surface to the dew-point temperature, creating saturated conditions and fog. This fog occurs especially on clear nights over moist ground.

Fog occurs in greatest frequency along the coastlines of the Pacific, the North Atlantic and the Labrador Sea, with over 80, 80–180, and 90–150 days of fog per year in these respective locations. This is due to cold air temperatures causing evaporation fog in the Atlantic and Labrador regions. While advection fog, the movement of warm air along a cold body of water, occurs along the Pacific, due to the cool temperature of the California Current. Advection fog also occurs along the Great Lakes, as warm air from the tropics travels over the Great Lakes giving regions such as Montreal more than 60 days of fog. Other inland locations receive a high number of days with fog, mostly associated with orographic or upslope fog where the warm, moist air is forced upslope and cools to create fog. Good examples of this are in the Cascades, causing 100–200 days of fog in Oregon and Washington states, and causing more than 80 days of fog along the Appalachian Mountains.

• *Describe* **air masses that affect North America and** *identify* **four types of atmospheric lifting mechanisms.**

22. How does a source region influence the type of air mass that forms over it? Give specific examples of each basic classification.

See Figure 5.24, p. 153, for details. Earth's surface imparts its varying characteristics to the air it touches. Such a distinctive body of air is called an air mass, and it initially reflects the characteristics of its source region. The longer an air mass remains stationary over a region, the more definite its physical attributes become. Within each air mass there is a homogeneity of temperature and humidity that sometimes extends through the lower half of the troposphere.

23. Of all the air masses, which are of greatest significance to the United States and Canada? What happens to them as they migrate to locations different from their source regions? Give an example of air-mass modification.

See Figure 5.24. As air masses migrate from their source regions, their temperature and moisture characteristics are modified. For example, an mT Gulf/Atlantic air mass may carry humidity to Chicago and on to Winnipeg, but gradually will become more stable with each day's passage northward. Similarly, temperatures below freezing occasionally reach into southern Texas and Florida, influenced by an invading winter cP air mass from the north. However, that air mass will have warmed from the –50°C (–58°F) of its source region in central Canada. In winter, as a cP air mass moves southward over warmer land, it is modified, warming especially after it crosses the snowline. If it passes over the warmer Great Lakes, it will absorb heat and moisture and produce heavy lake-effect snowfall downwind as far as New York,

Pennsylvania, New England, and the Maritime Provinces of Canada.

24. Explain why it is necessary for an air mass to rise if there is to be precipitation.

If air masses are to cool adiabatically (by expansion), and if they are to reach the dew-point temperature and saturate, condense, form clouds, and perhaps precipitate, then they must be lifted.

25. What are the four principal lifting mechanisms that cause air masses to ascend, cool, condense, form clouds, and perhaps produce precipitation? Briefly describe each.

The four principal lifting mechanisms causing adiabatic cooling are convergent lifting, convectional lifting, orographic lifting, and frontal lifting, (Figure 5.26). Convergent lifting occurs along the ITCZ, where air flows from areas of higher pressure towards areas of low pressure (pressure gradient force). The trade winds and easterlies form this type of convergence, warming air masses that create high vertical cumulonimbus clouds and high amounts of precipitation.

Convectional lifting is stimulated by local surface heating, often associated with differential heating between land and water surfaces. Maritime tropical air masses migrate out from the ITCZ transporting warm, moist unstable air to continental locations causing precipitation.

Adiabatic lifting occurs in both convergent and convectional lifting as warm, moist air masses rise vertically in the atmosphere, and are cooled by the surrounding air. Orographic lifting occurs when air masses are forced over barriers, such as mountains. Such movement causes the vertical movement of air masses, which are cooled by the cold air at higher elevations.

And frontal lifting takes place when air masses of differing temperature and humidity interact. Such collision is caused by the jet stream, affecting mostly midlatitude locations, as the jet stream pulls air masses from their source region into contact with other air masses. Adiabatic lifting occurs in frontal lifting when a cold air mass, or cold front, displaces warm air along the surface, or when a warm air mass, or warm front, slides over cooler air and is forced to rise vertically in the atmosphere.

26. Differentiate between the structure of a cold front and a warm front.

The leading edge of an advancing air mass is called its front. The leading edge of a cold air mass is a cold front, whereas the leading edge of a warm air mass is a warm front. A cold front is identified on weather maps by a line marked with triangular spikes pointing in the direction of frontal movement along an advancing cP air mass in winter. Warmer, moist air in advance of the cold front is lifted upward abruptly and is subjected to the same adiabatic rates and concepts of stability/instability that pertain to all lifting air parcels. A warm front is denoted on weather maps by a line marked with semicircles facing the direction of frontal movement. The leading edge of an advancing warm air mass is unable to displace cooler, passive air, which is denser. Instead, the warm air tends to push the cooler, underlying air into a characteristic wedge shape, with the warmer air sliding up over the cooler air.

Notes on Orographic Precipitation. See Figure 5.28. The physical presence of a mountain acts as a topographic barrier to migrating air masses. Orographic lifting occurs when air is pushed upslope against a mountain and cools adiabatically. Stable air that is forced upward in this manner may produce stratiform clouds, whereas unstable or conditionally unstable air usually forms a line of cumulus and cumulonimbus clouds. An orographic barrier enhances convectional activity and causes additional lifting during the passage of weather fronts and cyclonic systems, thereby extracting more moisture from passing air masses. The wetter intercepting slope is termed the windward slope, as opposed to the drier farside slope, known as the leeward slope, which is the location of a rain shadow. Moisture is condensed from the lifting air mass on the windward side of the mountain; on the leeward side the descending air mass is heated by compression, causing evaporation of any remaining water in the air. Thus, air can begin its ascent up a mountain warm and moist but finish its descent hot and dry.

Relative to the precipitation record on Mount Waialeale, Kauai mentioned in the text (annual average of 1234 cm, 486 in. or 40.5 ft), the rain gauge is at the 1547m (5075.5 ft) elevation and has been for the period of 1941 to the present. The previous record, one that is often cited in texts and articles, is for the 1931 to 1960 period—for a total of 1168 cm (460 in.) a year. The 1981.82 season received the highest precipitation for this station for a single 12-month period, a whopping 1585 cm (624 in.)! Check the updated publication about these and other records: Pauline Riordan and Paul G. Bourget, *World Weather Extremes*, ETL-0416. Fort Belvoir, VA: U.S. Army Corps of Engineers, Engineering Topographic Laboratories, December 1985 (updated previous edition dated January 1970).

You might want to obtain a copy of Climate of the States–Hawaii, Environmental Data Service, Washington: U.S. Government Printing Office; it includes a map of the islands, isohyet maps of each island, tables of data, and bibliography.

The record precipitation for contiguous North America falls to Canada, where the focusing effects of local topography and orographic lifting on the west side of Vancouver Island produced an average annual amount at Henderson Lake, British Columbia, of 650 cm (256 in.).

• *List* the measurable elements of weather and *describe* the life cycle of a midlatitude cyclone.

27. Differentiate between a cold front and a warm front as types of frontal lifting and what would you experience with each one.

Compare Figure 5.31, illustrating a cold front, and Figure 5.32, showing a warm front. The captions for each describe the differences.

28. How does a midlatitude cyclone act as a catalyst for conflict between air masses?

Wave cyclones form a dominant type of weather pattern in the middle and higher latitudes of both hemispheres and act as a catalyst for air-mass conflict as they bring contrasting air masses into conflict. A migrating center of low pressure, with converging, ascending air, spiraling inward counterclockwise in the Northern Hemisphere (or converging clockwise in the Southern Hemisphere), draws surrounding air masses into conflict in the cyclonic circulation along fronts.

29. Diagram a midlatitude cyclonic storm during its open stage. Label each of the components in your illustration, and add arrows to indicate wind patterns in the system.

See Figures 5.33 and 5.35.

30. What is your principal source of weather data, information, and forecasts? Where does your source obtain its data? Have you used the Internet and World Wide Web to obtain weather information? In what ways will you personally apply this knowledge in the future? What benefits do you see?

Personal analysis and response. The National Climate Data Center, of the National Environmental Satellite, Data, and Information Service, which is part of the National Oceanic and Atmospheric Administration, Department of Commerce, is located in Asheville, North Carolina 28801, and publishes a "Daily Weather Map" series on a weekly basis. This features a detailed surface map, 500-millibar chart, highest and lowest temperatures, and precipitation areas and amounts for the week. From the same data center, you can obtain a poster entitled "Explanation of the Daily Weather Map," which is a guide showing all the standard symbols presently used. These poster-charts are available in single copies (free) or in lots of 50 at a very low price.

A subscription to Weatherwise ("The Magazine About the Weather") is helpful, for it contains interesting articles, annual reviews of hurricanes and tornadoes, an annual weather photo contest, and historical information. It is published six times a year by the Helen Dwight Reid Foundation in association with the American Meteorological Society, Heldref Publications, Washington, D.C., 1-800-365-9753). Other periodicals of interest are the *Bulletin of the American Meteorological Society* (monthly), *Journal of Atmospheric Science*, and *Monthly Weather Review* from the American Meteorological Society; *Weather* (monthly) from the Royal Meteorological Society; and *NOAA* (bimonthly), from the Office of Public Affairs NOAA. (See the many links on our Internet Home Page.)

• *Analyze* **various types of violent weather—thunderstorms, tornadoes, and tropical cyclones.**

31. What constitutes a thunderstorm? What type of cloud is involved? What type of air mass would you expect in an area of thunderstorms in North America?

Tremendous energy is liberated by the condensation of large quantities of water vapor. This process is accompanied by violent updrafts and downdrafts. As a result, giant cumulonimbus clouds can create dramatic weather moments—squall lines of heavy precipitation, lightning, thunder, hail, blustery winds, and tornadoes. Thunderstorms may develop within an air mass, along a front (particularly a cold front), or where mountain slopes cause orographic lifting. Important here are the mT air masses of the Gulf and Atlantic source region.

32. Lightning and thunder are powerful phenomena in nature. Briefly describe how they develop.

Lightning refers to flashes of light caused by enormous electrical discharges—tens to hundreds of millions of volts—which briefly ignite the air to temperatures of 15,000° to 30,000°C (27,000° to 54,000°F). The violent expansion of this abruptly heated air sends shock waves through the atmosphere creating the sonic bangs known as thunder. The greater the distance a lightning stroke travels, the longer the thunder echoes. Lightning is created by a buildup of electrical energy between areas within a cumulonimbus cloud or between the cloud and the ground, with sufficient electrical potential to overcome the resistance of the atmosphere and leap from one surface to the other—it is like a giant spark.

A rule of thumb to use in determining the distance a lightning strike is from your location assumes that the flash arrived instantaneously at the speed of light, whereas the sound traveled at the speed of sound, some 3 seconds per km (1090 ft per sec, or 5 sec per mile). Simply begin counting at the moment of the flash to determine the elapsed time before you hear the thunder. Given the number of seconds elapsed you will know the distance you are from the lightning in km, feet, or mile units given above speeds. Suffice it to say that if you experience no delay and witness a simultaneous flash and crack of thunder then you are in the wrong place at the wrong time!

Thunder is enhanced by greater moisture density within the cloud and by topography that can act further to reverberate the sound waves.

33. Describe the formation process of a mesocyclone. How is this development associated with that of a tornado?

The updrafts associated with a cumulonimbus cloud appear on satellite images as pulsing bubbles of clouds. Because winds in the troposphere blow stronger above Earth's surface than they do at the surface, a body of air pushes forward faster at altitude than at the surface, thus creating a rotation in the air along a horizontal axis that is parallel to the ground. When that rotating air encounters the strong updrafts associated with frontal activity, the axis of rotation is shifted to a vertical alignment, perpendicular to the ground. It is this spinning, cyclonic, rising column of mid-troposphere-level air that forms a mesocyclone.

A mesocyclone can range up to 10 km (6 mi) in diameter and rotate over thousands of feet vertically within the parent cloud. As a mesocyclone extends vertically and contracts horizontally, wind speeds accelerate in an inward vortex (much as ice skaters accelerate while spinning by pulling their arms in closer to their bodies). A well-developed mesocyclone most certainly will produce heavy rain, large hail, blustery winds, and lightning; some mature mesocyclones will generate tornado activity.

34. Evaluate the pattern of tornado activity in the United States. What generalizations can you make about the distribution and timing of tornadoes?

See Figure 5.39. Of the 50 states, 49 have experienced tornadoes, as have all the Canadian provinces. May is the peak month. A small number of tornadoes are reported in other countries each year, but North America receives the greatest share because its latitudinal position and shape permit conflicting and contrasting air masses to have access to each other. Most tornadoes in the United States occur along a four-state corridor called tornado alley that includes Texas, Oklahoma, Kansas, and Nebraska.

Generally, tornadoes occur in conjunction with a mesocyclone, characterized by large cumulonimbus clouds, hail, intense winds, and lightning. Timing of tornadoes is still rather unpredictable, yet research with satellites, airplanes, and surface measurements is now enabling us to predict the occurrence of tornadoes. With the use of Doppler radar, which can detect the specific flow of moisture in mesocyclones, only 15% of American tornadoes strike without warning. Doppler radar enables forecasters to give tornado warnings 30 minutes to one hour in advance of an oncoming storm (See Figure 5.3.1, p. 164.).

Even though this text is being written in central California, far from the "tornado alley" of Oklahoma and Kansas, a tornado touched down just a block from this word processor on March 22, 1983. It was related to the incredible weather of the last intense El Niño phenomena—discussed in a Focus Study with Chapter 6. We were not at home and so no photos are included. It hit about 2 P.M., was moderate on the Fujita Scale, and moved northeastward, hopping along and damaging about 30 homes and several businesses.

For an analysis of tornadoes in Canada see M.J. Newark, "Tornadoes in Canada for the Period 1950 to 1979," CL 1-2-83, Downsview, Ontario: Analysis and Impact Division, Canadian Climate Centre, Atmospheric Environment Service, Environment Canada, 1983.

Note the photo in Figure 5.37. The May 1999 F5 tornado that hit urban sections of Oklahoma City was devastating and may have set some wind-speed and size records, since the funnel was almost a mile wide and persisted for more than 45 minutes.

35. What are the different classifications for tropical cyclones? List the various names used worldwide for hurricanes.

See Table 5.1 and Figure 5.40.

36. What forecast factors did scientists use to accurately predict the 1995 and other Atlantic hurricane seasons? Are there any trends in activity since 1995? How did the 2005 Atlantic tropical cyclone season compare? Please explain.

The IRM for the *Elemental Geosystems* fourth edition discussed the damage caused by Hurricane Andrew, which was a devastating storm. It was unusual in its intensity and was one of only a few category five hurricanes in the 20th century.

However, the 2005 Atlantic hurricane season set new records for the number and intensity of storms. There were 27 named storms and 14 hurricanes. Forecasters at the National Hurricane Center (NHC) in Miami were busy throughout the 1995–2005 hurricane seasons. This was the most active 11-year period in the history of the NHC, with 154 named tropical storms, including 88 hurricanes (37 intense, meaning category 3 or higher). This was a record level of activity despite the reduced number of tropical cyclones during the

1997 El Niño season of reduced tropical storm intensity. In 2003, the Atlantic tropical storm season was 168% of expected net tropical cyclone activity; 2004 hit 229% of the expected. For comparison, tropical storm activity for the 1995 season, discussed shortly, was 235% of an average season; the 2005 season with 22 named storms now is the leader.

An important innovation assisting the NHC and state and local governments with storm analysis is a new forecasting method developed by a team led by William M. Gray at Colorado State University. (See **http://typhoon.atmos.colostate.edu/forecasts/**.)

37. Relative to improving weather forecasting, what are some of the technological innovations discussed in this chapter?

The introductory section describes the new National Weather Service installations and approach: the Automated Surface Observing System (ASOS) arrays for gathering data; the new WSR-88D Doppler radar systems at 118 sites; and, the Advanced Weather Interactive Processing System (AWIPS) workstations that will be at 148 stations (pictured in Figures 1 and 2, News Report 5.3, pp. 164-5). This information places *Elemental Geosystems* at the forefront of contemporary reporting on the NWS.

Overhead Transparencies

89.	5.2 a, b	Land and water hemispheres
90.	5.3	Ocean and freshwater distribution, include oceans table
91.	5.4	Three states of water and water's phase changes
92.	5.6	Water's heat-energy characteristics
93.	5.7 and 5.8 a, b	Water vapor content and capacity (top); Dew-point temperature example (bottom)
94.	5.9	Satellite image water vapor full hemisphere
95.	5.10	Daily relative humidity charts
96.	5.11 and 5.12	Saturation vapor pressure; maximum specific humidity
97.	5.13 a, b	Hair hygrometer (top); sling psychrometer (bottom)
98.	5.14	Forces acting on an air parcel
99.	5.15 a, b	Adiabatic cooling and heating
100.	5.16	Temperature relationships and atmospheric stability
101.	5.17 a, b,	Stability two examples
102.	5.18	Principal clouds
103.	5.19 a, b	Cumulonimbus thunderhead
104.	5.20, 5.21, 5.22, 5.23	Fog types: advection, evaporation, valley, and radiation
105.	5.24 a and b	Principal air masses, winter and summer patterns
106.	5.25	Weather map for 2/17/01: cP air mass and high pressure
107.	5.26 a, b, c, d	Four atmospheric lifting mechanisms
108.	5.27	Local heating and convection
109.	5.28	Convectional activity over Florida; with locator map
110.	5.29 a, b	Orographic barrier produces precipitation (top); Sierra Nevada image and aerial photo (bottom)
111.	5.30 a, b, c	Cold front art (a), photo (b), and weather map (c)
112.	5.31	Warm front
113.	5.32 a b, c, and d	Idealized stages of a midlatitude cyclone; with legend of weather symbols
114.	5.33a, b	Typical and actual cyclonic storm tracks for North America
115.	5.34 a, b	SeaWiFS image for 4/20/00 (top); weather map segment (bottom)
116.	NR 5.3.2 a, b, c, d, e	ASOS, 3-D weather model, Advanced Workstation display, AWIPS work station, high-performance computing system
117.	5.35	Thunderstorm occurrence
118.	5.36 a, b, c	LIS lightning image for winter 2000 (top); and summer 2000 (middle); lightning photo (bottom)
119.	5.37 a, b, c	Mesocyclone and tornado formation
120.	5.38 a, b	Average tornadoes per year per 10,000 mi^2 (top); average tornadoes per month (bottom)
121.	5.39 and 5.28	Worldwide map of tropical cyclones
122.	5.40 a, b, c	Profile of a hurricane
123.	5.41 a, b	Hurricane Andrew, August 1992
124.	F.S. 5.1.1	1995 Atlantic hurricane season
125.	F.S. 5.1.2	Satellite observes four storms at once

6

Water Resources

This chapter begins by examining the water balance, which is an accounting of the hydrologic cycle for a specific area with emphasis on plants and soil moisture. We discuss the nature of groundwater and look at several examples of this generally abused resource. Groundwater resources are closely tied to surface-water budgets. Of course the ultimate repository of water on Earth is in the oceans. We also consider the water we withdraw and consume from available resources, in terms of both quantity and quality. The chapter addresses troubling prospects for the world's freshwater.

To supplement your materials for water resources please get a copy of a first-ever supplement to *National Geographic*: *Water—The Power, Promise, and Turmoil of North America's Fresh Water*, A National Geographic Special Edition, v. 184, no. 5A, 1993. Page 120 of this supplement presents an entire page of many resources that are useful in teaching. Sections include: "Supply—Sharing the Wealth of Water, and California: Desert in Disguise"; "Development—When Human's Harness Nature's Forces, and James Bay: Where Two Worlds Collide"; "Pollution—Troubled Waters Run Deep, and The Mississippi River Under Siege"; "Restoration—New Ideas, New Understanding, New Hope." The issue includes a giant map supplement of water resources, surface and ground, for the United States. Call 1-800-638-4077 to order this publication.

Outline Headings and Key Terms

The first-, second-, and third-order headings that divide Chapter 6 serve as an outline for your notes dies. The key terms and concepts that appear e in the text are listed here under their iate heading in **bold italics**. All these ted terms appear in the text glossary. Note the f box (❑) so you can mark your progress as

you master each concept. Your students have this same outline in their *Student Study Guide*. The ✪ icon indicates that there is an accompanying animation or other resource on the CD.

The outline headings for Chapter 6:
The Hydrologic Cycle
✪ **Earth's Water and the Hydrologic Cycle**
   ❑ *hydrologic cycle*
A Hydrologic Cycle Model
Surface Water
   ❑ *infiltration*
   ❑ *percolation*
Soil-Water Budget Concept
✪ **Global Water Balance Components**
   ❑ *soil-water budget*
The Soil-Water–Balance Equation
   Precipitation (PRECIP) Input
   ❑ *precipitation*
   ❑ *rain gauge*
   Actual Evapotranspiration (ACTET)
   ❑ *evaporation*
   ❑ *transpiration*
   ❑ *evapotranspiration*
   ❑ *potential evapotranspiration*
   Determining POTET
   ❑ *evaporation pan*
   ❑ *lysimeter*
Deficit
   ❑ *deficit*
   ❑ *actual evapotranspiration*
Surplus
   ❑ *surplus*
   ❑ *total runoff*
Soil Moisture Storage
   ❑ *soil moisture storage*
   ❑ *wilting point*
   ❑ *capillary water*
   ❑ *available water*
   ❑ *field capacity*
   ❑ *gravitational water*

The URLs related to this chapter of *Elemental Geosystems* can be found at
http://www.prenticehall.com/christopherson

Key Learning Concepts

After reading the chapter and using the Student Study Guide, the student should be able to:

• *Illustrate* the hydrologic cycle with a simple sketch and *label* it with definitions for each water pathway.
• *Relate* the importance of the water-budget concept to an understanding of the hydrologic cycle, water resources, and soil moisture for a specific location.
• *Construct* the water-balance equation as a way of accounting for the expenditures of water supply and *define* each of the components in the equation and its specific operation.
• *Describe* the nature of groundwater and *define* the elements of the groundwater environment.
• *Identify* critical aspects of freshwater supplies for the future and *cite* specific issues related to sectors of use, regions and countries, and potential remedies for any shortfalls.

Annotated Chapter Review Questions

• *Illustrate* **the hydrologic cycle with a simple sketch and** *label* **it with definitions for each water pathway.**

1. Sketch and explain a simplified model of the complex flows of water on Earth—the hydrologic cycle.

Vast currents of water, water vapor, ice, and energy are flowing about us continuously in an elaborate open global plumbing system. A simplified model of this complex system is useful to our study of the hydrologic cycle (Figure 6.1). The ocean provides a starting point, where more than 97% of all water is located and most evaporation and precipitation occur. If we assume that mean annual global evaporation equals 100 units, we can trace 86 of them to the ocean. The other 14 units come from the land, including water moving from the soil into plant roots and passing through their leaves. Of the ocean's evaporated 86 units, 66 combine with 12 advected from the land to produce the 78 units of precipitation that fall back into the ocean. The remaining 20 units of moisture evaporated from the ocean, plus 2 units of land-derived moisture, produce the 22 units of precipitation that fall over land. Clearly, the bulk of continental precipitation derives from the oceanic portion of the cycle.

2. What are the possible routes that a raindrop may take on its way to and into the soil surface?

Precipitation that reaches Earth's surface follows a variety of pathways. The process of precipitation striking vegetation or other ground cover is called interception. Precipitation that falls directly the ground, coupled with drips onto the ground vegetation, constitutes throughfall. Intercepted that drains across plant leaves and down plant termed stem flow and can represent an in moisture route to the surface. Water reaches the subsurface through infiltration, or penetration of the soil surface. It then permeates soil or rock through vertical movement called percolation.

3. Compare precipitation and evaporation volumes from the ocean with those over land. Describe advection flows of moisture and countering surface and subsurface runoff.

More than 97% of Earth's water is in the ocean, and here most evaporation and precipitation occur. 86% of all evaporation can be traced to the ocean. The other 14% comes from the land, including water moving from the soil into plant roots and passing through their leaves by transpiration. Of the ocean's evaporated 86%, 66% combines with 12% advected from the land to produce the 78% of all precipitation that falls back into the ocean. The remaining 20% of moisture evaporated from the ocean, plus 2% of land-derived moisture, produces the 22% of all precipitation that falls over land.

───────────

• *Relate* **the importance of the water-budget concept to an understanding of the hydrologic cycle, water resources, and soil moisture for a specific location.**

4. How might an understanding of the hydrologic cycle in a particular locale, or a soil-moisture budget of a site, assist you in assessing water resources? Give some specific examples.

A soil-moisture budget can be established for any area of Earth's surface by measuring the precipitation input and its distribution to satisfy the "demands" of plants, evaporation, and soil moisture storage in the area considered. A budget can be constructed for any time frame, from minutes to years. See Figures 6.10 and 6.11 for a specific examples.

───────────

• *Construct* **the water-balance equation as a way of accounting for the expenditures of water supply and *define* each of the components in the equation and their specific operation.**

5. Discuss the meaning of this statement "The soil-water budget is an assessment of the hydrologic cycle at a specific site."

A water balance can be established for any area of Earth's surface by calculating the total and station input and the total of various outputs. The balance approach allows an examination of the appropriate cycle, including estimation of streamflow at highligh c site or area, for any period of time. The check-o of the water balance is to describe the various which the water supply is expended. The water

balance is a method by which we can account for the hydrologic cycle of a specific area, with emphasis on plants and soil moisture.

Notes on Thornthwaite. The Thornthwaite method for estimating potential evapotranspiration (POTET) is still popular because of the ease of obtaining input data and therefore its usefulness for water analyses over large geographic areas. In fact, a water balance was actually calculated for the entire Earth by T.E.A. van Hylckama in 1956 when he took a composite of thousands of sites and regions (*The Water Balance of the Earth*, Laboratory of Climatology, Vol. 9, No. 2, 1956). Let us confine our approach to a smaller unit area.

The hydrologic cycle is a general portrait of Earth's water-flow system, whereas the water balance model allows us to examine the hydrologic cycle components in detail at a specific site for any unit of time.

The essential publication for this topic is Thornthwaite, Charles W., and John R. Mather. *The Water Balance.* Publications in Climatology, vol. 1. Centerton, NJ: Drexel Institute of Technology, Laboratory of Climatology, 1955. For additional overview and background, see the "Water Budget Analysis" entry in Oliver, John E., and Rhodes W. Fairbridge, eds. *The Encyclopedia of Climatology.* New York: Van Nostrand Reinhold Co., 1987. This is a good general reference for Chapters 5–7.

6. What are the components of the water-balance equation? Construct the equation and place each term's definition below its abbreviation in the equation.

See Figure 6.2. One approach is to obtain the supply and demand data for your area in a dry year and a wet year for contrast and comparison. For more advanced students and labs, you can work from scratch with mean monthly air temperature, latitude, and estimates of soil moisture capacity for various rooting depths. See Thornthwaite, Charles W., and John R. Mather. *Instructions and Tables for Computing Potential Evapotranspiration and the Water Balance.* Publications in Climatology, Vol. 3. Centerton, NJ: Drexel Institute of Technology, Laboratory of Climatology, 1957.

A complete list of Publications in Climatology, and of the publications mentioned here in particular, is available from C. W. Thornthwaite Associates, Route 1, Elmer, New Jersey 08318.

Thornthwaite and Mather's great contribution is the concept of POTET and an easy method for its estimation from generally available data. Thornthwaite prepared a volume of tables to facilitate the calculation procedure, although time-saving computer formats are available today. See Mather, John R., editor. *The*

Measurement of Potential Evapotranspiration, Vol. 7, No. 1. Seabrook, New Jersey: Johns Hopkins University, 1954.

7. Compare the annual water-balance data for Kingsport, Tennessee; Ottawa, Ontario; and Phoenix, Arizona. What differences do you see in the components? What similarities? What does this tell you about local water resources in each area?

Students can use data from Figure 6.9, 6.10, and on our Home Page on the Internet (climate data under Chapter 7 in main menu for the companion text *Elemental Geosystems* 5/E), to perform the basic PRECIP–POTET bookkeeping (water accounting) procedure to determine net supply or demand. You will need to assign values for soil moisture storage such as the 100 mm (4.0 in.) used in Figure 6.10.

Comparing PRECIP with POTET by month determines whether there is a net supply (+) or net demand (–) for water. We can see that Kingsport experiences a net supply from January through May and from October through December. However, the warm days and months from June through September result in net demands for water.

8. Explain how to derive actual evapotranspiration (ACTET) in the water-balance equation.

The actual amount of evaporation and transpiration that occurs is derived by subtracting DEFIC, or water demand, from POTET. Under ideal conditions, POTET and ACTET are about the same, so that plants do not experience a water shortage. Droughts result from deficit conditions, where ACTET is greater than the available moisture.

9. What is potential evapotranspiration (POTET)? How do we go about estimating this potential rate? What factors did Thornthwaite use to determine this value?

POTET is the amount of moisture that would evaporate and transpire if the moisture were available; the amount lost under optimum moisture conditions— the moisture demand. Both evaporation and transpiration directly respond to climatic conditions of temperature and humidity. For the empirical measurement of POTET, probably the easiest method employs an evaporation pan, or evaporimeter. As evaporation occurs, water in measured amounts is automatically replaced in the pan. Screens of various-sized mesh are used to protect against overmeasurements created by wind. A lysimeter is a relatively elaborate device for measuring POTET, for an actual portion of a field is isolated so that the moisture moving through it can be measured. See Figure 6.5 for a schematic of such a device.

Thornthwaite found that POTET is best approximated using mean monthly air temperature (measured with a thermometer) and daylength (a function of the measuring station's latitude). Both factors are easily determined, making it relatively simple to calculate POTET and the other water-balance components in the equation indirectly, with fair accuracy for most midlatitude locations.

10. Explain the operation of soil-moisture storage, soil-moisture utilization, and soil-moisture recharge. Include discussion of field capacity, capillary water, and wilting point concepts.

Soil moisture storage (ΔSTRGE) refers to the amount of water that is stored in the soil and is accessible to plant roots, or the effective rooting depth of plants in a specific soil. This water is held in the soil against the pull of gravity and is discussed in more detail in Chapter 15. Soil is said to be at the wilting point when plant roots are unable to extract water; in other words, plants will wilt and eventually die after prolonged moisture deficit stress.

The soil moisture that is generally accessible to plant roots is capillary water, held in the soil by surface tension and cohesive forces between the water and the soil. Almost all capillary water is available water in soil moisture storage and is removable for POTET demands through the action of plant roots and surface evaporation; some capillary water remains adhered to soil particles along with hygroscopic water. When capillary water is full in a particular soil, that soil is said to be at field capacity, an amount determined by actual soil surveys.

When soil moisture is at field capacity, plant roots are able to obtain water with less effort, and water is thus rapidly available to them. As the soil water is reduced by soil moisture utilization, the plants must exert greater effort to extract the same amount of moisture. Whether naturally occurring or artificially applied, water infiltrates soil and replenishes available water content, a process known as soil moisture recharge.

Notes on Deficit. Thornthwaite and Mather define drought at several different levels of intensity. The more arid areas of the west experience *permanent drought* measured in chronic deficits. During the year, some areas experience *seasonal drought* as patterns of precipitation occur in predictable rainy and dry seasons. *Irregular droughts* occur within the normal variability of precipitation from time to time. And finally, *invisible drought* is defined as an episode when all of POTET is not satisfied even though precipitation is occurring. Invisible drought is caused by the inefficiency of soil moisture removal and use by plants when the soil is less than field capacity. This subtle, almost unseen, drought can be just as damaging to plants and crop yields as visible drought.

NOAA and the National Weather Service publish weekly Palmer Crop Moisture Index reports that give soil moisture conditions for the United States. (See the "Weekly Weather and Crop Bulletin" published jointly by the U.S. Department of Agriculture and the Environmental Data Service of NOAA, and available in single copies or by subscription. It is also available on the web at **http://www.drought.noaa.gov/index.html.** The Palmer Crop Moisture Index is overviewed by its author, Wayne C. Palmer in, "Keeping Track of Crop Moisture Conditions...The New Crop Moisture Index." *Weatherwise* (August 1968): pp. 157–162.)

The Palmer system regards drought as a significant reduction in soil moisture below plant requirements. Palmer established a Drought Severity Index to rate water balance conditions. Severity ratings range from mild (-1), moderate (-2), severe (-3), to extreme (-4). These drought severity indices have measured as far back as A.D. 1700 through tree-ring analysis and demonstrate an interesting periodicity for midlatitude droughts in the American West and Midwest that coincide with the Hale Sunspot Cycle.

By comparing precipitation and soil moisture with POTET, a much more accurate estimation of drought is possible than with the traditional method of reporting "days without rain." Deficit avoidance provides a useful basis for the management of irrigation applications, so that through careful management not only can obvious drought be deduced for a large area, but small changes in moisture availability to crops also can be estimated. Added irrigation water, if available, augments precipitation, is considered an input to the water balance calculation, and can be handled much as one would a simple bookkeeping procedure.

The drought of 1988 is a recent case in point. Record temperatures for this century took place over much of the United States, whereas precipitation was almost nonexistent except for some relief late in the summer. Most of the United States in 1988 recorded high POTET and low PRECIP and therefore experienced higher than normal soil moisture utilization. Many areas had deficits (unmet POTET) and eventually widespread soil moisture shortages from the Mississippi River to Colorado, and from Lake Tahoe to Texas. Some of the late summer soil moisture recharge that did take place was simply too late to benefit the crops.

11. In the case of silt-loam soil from Figure 6.8, roughly, what is the available water capacity? How is this value derived?

See Figure 6.8 and the explanation of soil moisture in the last question. The lower line on the graph plots the wilting point; the upper line plots field capacity. The space between the two lines represents the amount of water available to plants given varying soil textures. Different plant types growing in various types of soil send roots to different depths and therefore are exposed to varying amounts of soil moisture. For example, shallow-rooted crops such as spinach, beans, and carrots send roots down 65 cm (25 in.) in a silt loam, whereas deep-rooted crops such as alfalfa and shrubs exceed a depth of 125 cm (50 in.) in such a soil. A soil blend that maximizes available water is best for supplying plant water needs.

12. For water resources, what does it mean to move water geographically or across the calendar?

Reservoirs and water-retention facilities can hold back runoff from a rainy season for release later in the year in a drier season—this moves water across the calendar. In some regions, such as California, the population and greater water demand (southern California) do not match the water supply (northern California); or in Australia the rainy southeast coast is in contrast to the dry interior (see next question). Moving water geographically from surplus regions to regions of deficit constitutes a geographic realignment of water resources.

13. In terms of water balance and water management, explain the logic behind the Snowy Mountain Scheme in southeastern Australia.

In the Snowy Mountains, part of the Great Dividing Range in extreme southeastern Australia, precipitation ranges from 100 to 200 cm (40 to 80 in.) a year, whereas interior Australia receives under 50 cm (20 in.), and drops to less than 25 cm (10 in.) further inland. POTET values are high throughout the Australian outback and lower in the higher elevations of the Snowy Mountains. The plan was designed to take surplus water that flowed down the Snowy River eastward to the Tasman Sea and reverse the flow to support newly irrigated farmland in the interior of New South Wales and Victoria. The westward flow of the Murray, Tumut, and Murrumbidgee rivers is augmented, and as a result, new acreage is now in production in what was dry outback, formerly served only by wells drawing on meager groundwater supplies.

• *Describe* the nature of groundwater and *define* the elements of the groundwater environment.

14. Are groundwater resources independent of surface supplies, or are the two interrelated? Explain your answer.

Groundwater is the part of the hydrologic cycle that lies beneath the ground and is therefore tied to surface supplies. Groundwater is the largest potential source of freshwater in the hydrologic cycle—larger than all surface reservoirs, lakes, rivers, and streams combined. Between Earth's surface and a depth of 3 km (10,000 ft) worldwide, some 8,340,000 km^3 (2,000,000 mi^3) of water resides.

15. Make a simple sketch of the subsurface environment, labeling zones of aeration and saturation and the water table in an unconfined aquifer. Then add a confined aquifer to the sketch.

Utilize Figure 6.14 as the basis for this sketch.

16. At what point does groundwater utilization become groundwater mining? Use the High Plains aquifer example to explain your answer.

Aquifers frequently are pumped beyond their flow and recharge capacities; groundwater mining refers to this over utilization of groundwater resources. Large tracts of the Midwest, West, lower Mississippi Valley, and Florida experience chronic groundwater overdrafts. In many places the water table or artesian water level has declined more than 12 m (40 ft). Groundwater mining is of special concern today in the High Plains aquifer, which is the topic of Focus Study 6.2.

17. What is the nature of groundwater pollution? Can contaminated groundwater be cleaned up easily? Explain.

When surface water is polluted, groundwater also becomes contaminated because it is fed and recharged from surface water supplies. Groundwater migrates very slowly compared with surface water. Surface water flows rapidly and flushes pollution downstream, but sluggish groundwater, once contaminated, remains polluted virtually forever. Pollution can enter groundwater from industrial injection wells, septic tank outflows, seepage from hazardous-waste disposal sites, industrial toxic-waste dumps, residues of agricultural pesticides, herbicides, fertilizers, and residential and urban wastes in landfills. Thus, pollution can come either from a point source or from a large general area (a non-point source), and it can spread over a great distance, as illustrated in several ways in Figure 6.14. Because surface water flows rapidly, it can flush pollution downstream. Yet if groundwater is polluted, because it is slow moving, once it's contaminated, it will remain polluted virtually forever.

• *Identify* critical aspects of freshwater supplies for the future and *cite* specific issues related to sectors of use, regions and countries, and potential remedies for any shortfalls.

18. Describe the principal pathways involved in the water budget of the contiguous 48 states. What is the difference between withdrawal and consumptive use of water resources? Compare these with instream uses.

The 4200 BGD that makes up the U.S. water supply is not evenly distributed across the country or during the calendar year. Figure 6.17 shows that the two principal routes of expenditure of the daily water supply are evapotranspiration from non-irrigated land and surplus. About 71% (2970 BGD) of the daily supply passes through non-irrigated land—farm crops and pastures; forest and browse (young leaves, twigs, and shoots); and non-economic vegetation–and eventually returns to the atmosphere for another journey through the hydrologic cycle. Only 29%, or 1231 BGD, becomes surplus runoff, available for withdrawal and consumption.

Consumptive uses are those that remove water from the budget at some point, without returning it farther downstream. Withdrawal, or non-consumptive use, refers to water that is removed from the supply and used for various purposes, then returned to the same supply. Instream uses are those that use stream water in place; navigation, fishing, hydroelectric power, and ecosystem preservation.

19. Characterize each of the sectors withdrawing water: irrigation, industry, and municipalities. What are the present usage trends in developed and less developed nations?

Figure 6.18 illustrates the three main flows of this withdrawn water in the United States: irrigation (34%), industry (57%), and municipalities (9%). In contrast, Canada uses only 7% of its withdrawn water for irrigation, and 84% for industry. A sector analysis for global water resources is in Figure 6.19. About one-fourth of the annual renewable water worldwide will be actively utilized by the year 2000. By that time, one-half billion people will be depending on the polluted Ganges River alone. Furthermore, of the 200 largest river basins in the world (basins where rivers drain into an ocean, lake, or inland sea), 148 are shared by two nations, and 52 are shared by from three to ten nations. Clearly, water issues are international in scope, yet we continue toward a water crisis without a concept of a world water economy as a frame of reference.

20. Briefly assess the status of world water resources. What challenges are there in meeting future needs of an expanding population and growing economies? How do you think world-water-economy thinking and planning might begin?

About 30% of the annual renewable water worldwide is presently utilized. The World Bank estimates that $600 billion in capital investment is needed by 2010 to augment current water resources. In addition, aspects of global climate change will worsen water supply conditions for many regions and therefore increase this cost estimate. Available water declines as population increases, and individual demand increases with economic development. Figure 6.20 portrays a 1997 assessment for 1995 of global water scarcity. Freshwater is now scarce in many regions of the world, and unlike other commodities such as oil or wheat, water has no substitute.

21. How does acid deposition (acid rain) tie together the issues of air pollution, energy production and consumption from Chapter 2 and Focus Study 2.2 and water resource quality in this chapter?

The principal sources for anthropogenic nitrogen dioxide and sulfur oxides in the environment are from the combustion of fossil fuels, namely coal and oil. The principal source for the primary radiatively active greenhouse gas, carbon dioxide, is the combustion of fossil fuels. The acidification of water resources by acid deposition (nitric and sulfuric acids) is thus linked to fossil fuel consumption. Therefore, any policies to reduce fossil fuel consumption through efficiency and conservation measures will, in turn, clean the air, reduce greenhouse warming rates, and reduce acids in waterways. Additionally, all such strategies save money for the consumer and reduce other health, medical, and environmental costs to the economy.

Overhead Transparencies

150.	7.1	The hydrologic cycle model
151.	7.2a, b	(b) Soil-moisture environment (top); (a) Water balance equation and explanation equation (bottom)
152.	7.4, 7.6	Precipitation in the United States and Canada (top); and potential evapotranspiration demand (bottom)
153.	7.5	Weighing lysimeter
154.	7.7 and 7.8	Types of soil moisture (top); soil-moisture availability (bottom)
155.	7.9	Sample water budget—Kingsport, Tennessee
156.	7.10a, b	Sample water balance stations
157.	7.11	Annual global river runoff
158.	7.12	Groundwater resource potential
159.	7.13 (left)	Groundwater characteristics
160.	7.13 (right)	Groundwater characteristics, cont.
161.	7.15a, b	Influent and effluent stream
162.	7.17	U.S. annual water budget
163.	7.18 and 7.19	Water withdrawal by sector for world (top); global water scarcity (bottom)
164.	F.S. 7.1.2 a, b	Hurricane Camille attributable precipitation (top); deficit abatement (bottom)
165.	F.S. 7.2-1 a, b	Map of High Plains Aquifer (left); water-level changes 1980–1995 (right)

Global Climate Systems

Earth experiences an almost infinite variety of weather. Even the same location may go through periods of changing weather. This variability, when considered along with the average conditions at a place over time, constitutes climate. Climates are so diverse that no two places on Earth's surface experience exactly the same climatic conditions, although general similarities permit grouping and classification. Chapter 7 serves as a synthesis of content from Parts One and Two of the text.

Finally, we take a look at how Earth's temperature system appears to be in a state of dynamic change as concerns about climate change and global warming and potential episodes of global cooling are discussed. This chapter presents an overview of the climatic effects of these temperature changes.

Outline Headings and Key Terms

The first-, second-, and third-order headings that divide Chapter 7 serve as an outline for your notes and studies. The key terms and concepts that appear **boldface** in the text are listed here under their appropriate heading in ***bold italics***. All these highlighted terms appear in the text glossary. Note the check-off box (❑) so you can mark your progress as you master each concept. Your students have this same outline in their *Student Study Guide*. The ✍ icon indicates that there is an accompanying animation or other resource on the CD.

The outline headings for Chapter 7:
- ❑ *climate*

Earth's Climate System and Its Classification
- ✍ **Global Climates: genetic map (causal)**
- ✍ **Global Climates: empirical map (Köppen)**
- ✍ **Global Patterns of Precipitation**
  - ❑ *climatology*
  - ❑ *climatic regions*

Climate Components: Insolation, Temperature, Air Pressure, Air Masses, and Precipitation
Classification of Climatic Regions
- ❑ *classification*
- ❑ *genetic classification*
- ❑ *empirical classification*
A Climate Classification System
    Classification Categories
    Global Climate Patterns
- ❑ *climographs*

Tropical Climates (equatorial and tropical latitudes)
    Tropical Rain Forest Climates
    Tropical Monsoon Climates
    Tropical Savanna Climates
Mesothermal Climates (midlatitudes, mild winters)
    Humid Subtropical Hot Summer Climates
    Marine West Coast Climates
    Mediterranean Dry-Summer Climates
Microthermal Climates (mid and high latitude, cold winter)
    Humid Continental Hot Summer Climates
    Humid Continental Mild Summer Climates
    Subarctic Climates
Polar and Highland Climates
- ✍ **Geographic Scenes: East Greenland Photo Gallery**
- ✍ **High Latitude Animals Photo Gallery**
- ✍ **High Latitude Connection videos**
    Tundra Climate
    Ice Cap and Ice Sheet Climate
       Polar Marine Climate
Arid and Semiarid Climates (permanent moisture deficits)
    Desert Characteristics
    Low-Latitude Hot Desert Climates
    Midlatitude Cold Desert Climates
    Low-Latitude Hot Steppe Climates
    Midlatitude Cold Steppe Climates
Global Climate Change
- ✍ **Global warming, climate change**
    Global Warming

The URLs related to this chapter of *Elemental Geosystems* can be found at
http://www.prenticehall.com/christopherson

Key Learning Concepts

After reading the chapter and using the Student Study Guide, the student should be able to:

• *Define* climate and climatology and *explain* the difference between climate and weather.
• *Review* the role of temperature, precipitation, air pressure, air mass patterns, and sea-surface temperatures used to establish climatic regions.
• *Review* the development of climate classification systems and *compare* genetic and empirical ways of classifying climate.
• *Describe* the principal climate classification categories other than deserts and *locate* these regions on a world map.
• *Explain* the precipitation and moisture efficiency criteria used to determine the arid and semiarid climates and locate them on a world map.
• *Outline* future climate patterns from forecasts presented and *explain* the causes and potential consequences.

Annotated Chapter Review Questions

• *Define* climate and climatology and *explain* the difference between climate and weather.

1. Define climate and compare it with weather. What is climatology?

Earth experiences an almost infinite variety of weather. Even the same location may go through periods of changing weather. This variability, when considered along with the average conditions at a place over time, constitutes climate. In a traditional framework, early climatologists faced the challenge of identifying patterns as a basis for establishing climatic classifications. Currently at the forefront of scientific effort by climatologists are developing models that can simulate the vast interactions and causal relationships of the atmosphere and hydrosphere. Climatology, the study of climate, involves analysis of the patterns in time and space created by various physical factors in the environment.

For an interesting comparison of similar elements that evolved to produce extremely different climates, see Robert M. Haberle, "The Climate of Mars," *Scientific American* (May 1986): pp. 54–62. The landing of *Pathfinder* and the *Sojourner* rover on Mars gave JPL scientists the opportunity to give real-time weather reports from the landing site. A typical day there had temperatures ranging from $-14°C$ (7°F) to $-73°C$ ($-100°F$), with winds at 20 mph and atmospheric pressure of 6.74 mb at noon (equivalent air pressure at the 39 km (24 mi) altitude level in Earth's atmosphere. Contact: **http://www.JPL.nasa.gov**

2. Explain how a climatic region synthesizes climate statistics.

Weather observations, gathered simultaneously from different points within a region, are plotted on maps and are compared to identify climatic regions. The weather components that combine to produce climatic regions include insolation, temperature, humidity, seasonal precipitation, atmospheric pressure and winds, air masses, types of weather disturbances, and cloud coverage. Similar climatic regions experience many of the same weather components.

Climatic information can be found on the Internet. *The Climatic Data Catalogue* is accessible at: **http://rainbow.ldgo.columbia.edu**.
This site includes global atmospheric circulation statistics, Navy bathometry, and surface climatologies. Another good site is the Climate Diagnosis Center located at:
http://noaacdc.colorado.edu/cdu/cdc_home.html.
This site has weather statistics spanning centuries. Lastly, the Climatic Research Unit at the University of

East Anglia, England has information about current weather, climatic research and links to other climate research facilities around the world. Contact them at: **http://www.cru.uea.ac.uk**.

A necessary resource is *The Encyclopedia of Climatology* edited by John E. Oliver and Rhodes W. Fairbridge (New York: Van Nostrand Reinhold Co., 115 Fifth Avenue, New York, NY 10003, 1987). The section on climatic classification covers the basics of classification, empirical systems, the Köppen system, the Thornthwaite system, and genetic systems. This volume also contains extensive sections on climatic data, with a list of helpful references. Appropriate chapters in Glenn T. Trewartha's several classic textbooks contain valuable material on climates and classification modification that he established (four titles with McGraw-Hill). Glenn T. Trewartha's *The Earth's Problem Climates* (Madison: The University of Wisconsin, 1961) examines climates worldwide, particularly those that do not conveniently fit expected global patterns. Also refer to the chapter on climatic classification in Howard J. Critchfield's *General Climatology*, Englewood Cliffs: Prentice-Hall, 1966. He describes both the Köppen and Thornthwaite systems.

3. How does the El Niño phenomenon produce the largest interannual variability in climate? What are some of the changes and effects that occur world-wide?

Normally, as shown in Figure 4.21, p. 127, the region off the West coast of South America is dominated by the northward-flowing Peru Current. These cold waters move toward the equator and join the westward movement of the south equatorial current. The Peru current is part of the overall counter-clockwise circulation that normally guides the winds and surface ocean currents around the subtropical high-pressure cell dominating the eastern subtropical Pacific (visible on the world pressure maps in Figure 4.11, p. 116). Occasionally, and for unexplained reasons, pressure patterns alter and shift from their usual locations, thus affecting surface ocean currents and weather on both sides of the Pacific. Unusually high pressure develops in the western Pacific and lower pressure in the eastern Pacific. This regional change is an indication of large-scale ocean-atmosphere interactions. Trade winds normally moving from east to west weaken and can be replaced by an eastward (west-to-east) flow. Sea-surface temperatures off South America then rise above normal, sometimes becoming more than 8C° (14F°) warmer, replacing the normally cold, upwelling, nutrient-rich water along Peru's coastline.

• *Review* the role of temperature, precipitation, air pressure, air-mass patterns and sea-surface temperatures used to establish climatic regions.

4. How do radiation receipts, temperature, air pressure inputs, and precipitation patterns interact to produce climate types? Give an example from a humid environment and one from an arid environment.

Uneven insolation over Earth's surface, varying with latitude, is the energy input for the climate system (see Chapter 2). Daylength and temperature patterns vary diurnally and seasonally. Average temperatures and daylength are the basic factors that help us approximate POTET (potential evapotranspiration; portrayed for North America in Figure 7.6, p. 227). The principal controls of temperature are latitude, altitude, land-water heating differences, and the amount and duration of cloud cover. The pattern of world temperatures and annual temperature ranges are discussed in Chapter 3. The moisture input to climate is precipitation in the forms of rain, sleet, snow, and hail. Figure 7.4, p. 225, portrays the distribution of mean annual precipitation in North America; Figure 6.2 shows the worldwide distribution of precipitation and identifies several patterns.

5. Evaluate the relationships among a climatic region, ecosystem, and biome.

One type of climatic analysis involves discerning areas of similar weather statistics and grouping these into climatic regions that contain characteristic weather patterns. Climate classifications are an effort to formalize these patterns and determine their related implications to humans. Interacting populations of plants and animals in an area forms a community. An ecosystem involves the interplay between a community of plants and animals and its abiotic physical environment. A biome is a large, stable terrestrial ecosystem.

• *Review* the development of climate classification systems and *compare* genetic and empirical systems as ways of classifying climate.

6. What are the differences between a genetic and an empirical classification system?

Classification is the process of ordering or grouping data or phenomena in related classes. A classification based on causative factors—for example, the genesis of climate based on the interaction of air masses is called a genetic classification. An empirical classification is based on statistics or other data used to determine general categories. Climate classifications based on temperature and precipitation data are empirical classifications.

7. What are some of the climatological elements used in classifying climates?

Some of the key components in classifying climates are insolation, temperature, air pressure, air masses and precipitation. Uneven insolation over Earth's surface, varying with latitude, is the energy input for the climate system. Daylength and temperature patterns vary daily and seasonally (Chapter 2). The principal controls of temperature are latitude, altitude, land-water heating differences, and the amount and duration of cloud cover. The pattern of world temperatures and annual temperature ranges is in Chapter 3 and portrayed in Figures 3.24, 3.26, and 3.28.

The hydrologic cycle provides the basic means of transferring energy and mass through Earth's climate system. Figure 6.2 shows the worldwide distribution of precipitation (rain, sleet, snow, and hail). You can identify several patterns, such as the way precipitation decreases from Western Europe inland toward the heart of Asia (continentality). However, high precipitation values dominate southern Asia, where the landscape receives moisture-laden monsoonal winds from the Indian Ocean in the summer. Southern South America reflects a pattern of wet western (windward) slopes and dry eastern (leeward) slopes (orographic effects). (See Figure 7.4, which portrays the distribution of average annual precipitation in North America.) Figure 6.2 also shows worldwide average annual precipitation.

Most of Earth's deserts and regions of permanent drought are located in lands dominated by subtropical high-pressure cells, with bordering lands grading to grasslands and to forests as precipitation increases. The most consistently wet places on Earth straddle the equator in the Amazon region of South America, the Congo (Zaire) region of Africa, and the Indonesian and Southeast Asian area, all of which are influenced by equatorial low pressure and the intertropical convergence zone (Figures 4.11, 4.13).

This edition presents climate classification using a genetic approach, rather than Köppen's as in past editions. The new climate map, Figure 6.5, does not have the Köppen letter designations; however, it retains the descriptive names for Köppen's climates and now incorporates the causes of climates. For those instructors who wish to use the Köppen letter designations, Appendix A is a map of Köppen's climate zones with the Köppen criteria for each zone around the border. Both the Köppen map and Figure 6.5 are drawn using the same color scheme, so students can view the causative factors involved with climate, as well as the Köppen classifications.

Notes on Köppen: Wladimir Köppen (1846–1940), a German climatologist and botanist, designed the Köppen classification system, widely used for its ease of comprehension. First published in stages, his classification began with an article on heat zones in 1884. By 1900, he was considering plant communities in his selection of some temperature criteria, using a world vegetation map prepared by French plant physiologist A. de Candolle in 1855. Letter symbols then were added to designate climate types. Later he reduced the role played by plants in setting boundaries and moved his system strictly toward climatological empiricism. The first wall map showing world climates, coauthored with his student Rudolph Geiger, was introduced in 1928 and soon was widely used.

Please refer to Arthur A. Wilcox's "Köppen after Fifty Years." *Annals of the Association of American Geographers* 58, No. 1 (March 1968): 12–28, for it provides interesting background on Köppen and the development of his classification. Although it is an empirical classification, aspects of his early genetic considerations involving plants are discussed. He selected temperature criteria in an attempt to match vegetation patterns; his classification remains based on climatic data.

The Köppen system is best viewed for what it is: a valuable tool for general understanding, best limited to small scale hemispheric and world maps showing general climatic relationships and patterns. Most criticism of his system stems from asking the classification model to do what it was not designed to do, that is, produce specific climatic descriptions for local areas.

For arguments that the Thornthwaite system should be allowed to supersede the Köppen system, see Douglas B. Carter, "Farewell to the Köppen Classification of Climate" from Abstracts of the papers presented at the 63rd annual meeting of the Association of American Geographers, St. Louis, Missouri, 11–14 April 1967. Printed in the *Annals of the Association of American Geographers* Vol. 57, No. 4 (December 1967). Here is some background on the Thornthwaite system.

The development of a simple method for the determination of potential evapotranspiration led Thornthwaite to his climate classification system. Thornthwaite was critical of Köppen's choice of criteria for his climatic boundaries, and especially, the boundaries between the humid and dry climates. Temperature efficiency and precipitation effectiveness were concepts contributed by Thornthwaite. His 1948 classification introduced a moisture index concept as a basis for classification. Thornthwaite's classification is marred only by its complexity and lack of widespread use. Otherwise, the system in several ways is more accurate than is the Köppen system in its depiction of the humid-dry boundaries, especially those in North America.

The key to Thornthwaite's approach is that temperature and precipitation alone are not the most active factors in the distribution of vegetation; rather, Thornthwaite regarded POTET and its relation to precipitation and plant moisture needs as the critical factor. Whereas Köppen used average annual temperature and precipitation for the determination of a moisture index, Thornthwaite used POTET and established a moisture index based on calculations of water balance moisture surpluses and moisture deficits as discussed in Chapter 7. The moisture index can range from +100 as a measure of the degree PRECIP exceeds POTET, to a low of −100, where no PRECIP is received. When the moisture index is at zero, it is at the midpoint along the boundary line between the humid and dry climates. Thornthwaite established five climate classifications: arid, semiarid, subhumid, humid, and perhumid. A specific area to compare Thornthwaite with the Köppen world map is the transition region between the dry and humid climates from the western prairies to the Rocky Mountains. Köppen extended the humid climates, both C and D, west of the 100° W meridian, providing poor definition of the region. In reality, the region goes through a transition from tall grass to short grass to semiarid steppe to arid zones (see Chapter 16). On the Thornthwaite map, this transition area is clearly divided into moist subhumid, dry subhumid, and semiarid classifications, thereby providing greater clarification.

• *Describe* the principal climate classification categories other than deserts and *locate* these regions on a world map.

8. List and discuss each of the principal climate categories. In which one of these general types do you live? Which classification is the only type associated with the annual distribution and amount of precipitation?

See Figure 6.3 and 6.4 for the climatic classification system. Figure 6.5 is a climate map showing global influences on climate, including the distribution of air pressure, air mass source regions, and ocean currents.

9. What is a climograph, and how is it used to display climatic information?

Climographs for cities that exemplify particular climates are presented in the text. These climographs show monthly temperature and precipitation, location coordinates, annual temperature range, total annual precipitation, annual hours of sunshine (as an indication of cloudiness), the local population, and a location map.

10. Which of the major climate types occupies the most land and ocean area on Earth?

In terms of total land and ocean area, tropical climates are the most extensive, occupying about 36% of Earth's surface. The tropical climate classification extends along all equatorial latitudes, straddling the tropics from about 20° N to 20° S and stretching as far north as the tip of Florida and south-central Mexico, central India, and southeast Asia.

11. Characterize the *tropical* climates in terms of temperature, moisture, and location.

The key temperature criterion for a *tropical* climate is that the coolest month must be warmer than 18°C (64.4°F), making these climates truly winterless. The consistent daylength and almost perpendicular Sun angle throughout the year generates this warmth. Subdivisions of the *tropical* climates are based upon the distribution of precipitation during the year. Thus, in addition to consistent warmth, a *tropical rain forest* climate is constantly moist, with no month recording less than 6 cm (2.4 in.) of precipitation. Indeed, most stations in a tropical rain forest climate receive in excess of 250 cm (100 in.) of rainfall a year. Not surprisingly, the water balances in these regions exhibit enormous water surpluses, creating the world's largest stream discharges in the Amazon and Congo (Zaire) Rivers.

12. Using Africa's *tropical* climates as an example, characterize the climates produced by the seasonal shifting of the ITCZ with the high Sun.

Those areas that are covered by the ITCZ during all 12 months of the year constitute the tropical rain forest climates. As the ITCZ shifts with the high Sun, new areas are affected, whereas opposite latitudes of lower Sun angles are abandoned by the rains of the ITCZ. Generally, regions that are covered from 6 to 12 months (essentially along coastal areas) are tropical monsoon; those receiving coverage by the ITCZ less than 6 months of the year are principally the *tropical savanna* climatic regions.

13. *Mesothermal (subtropical and midlatitude, mild winter)* climates occupy the second largest portion of Earth's entire surface. Describe their temperature, moisture, and precipitation characteristics.

The word *mesothermal* suggests warm and temperate conditions, with the coldest month averaging below 18°C (64.4°F) but with all months averaging above 0°C (32°F). The *mesothermal* climates, and nearby portions of the *microthermal* climates, are regions of great weather variability, for these are the latitudes of greatest air-mass conflict. The *mesothermal* climatic region marks the beginning of true seasonality, with contrasts in temperature as evidenced by vegetation, soils, and human lifestyle adaptations. Subdivisions of the *mesothermal* classification are based on precipitation variability as given in Figure 6.5.

14. Explain the distribution of the *humid subtropical hot summer* and *Mediterranean dry-summer* climates at similar latitudes and the difference in precipitation patterns between the two types. Describe the difference in vegetation associated with these two climate types.

Mesothermal *humid subtropical hot summer* climates are located in the eastern and east-central portions of the continents and are influenced during the summer by the maritime tropical air masses generated over warm coastal waters off eastern coasts. The warm, moist, unstable air forms convectional showers over land. In fall, winter, and spring, maritime tropical and continental polar air masses interact, generating frontal activity and frequent midlatitude cyclonic storms.

Across the planet during summer months, shifting cells of subtropical high pressure block moisture-bearing winds from adjacent regions. As an example, in summer the continental tropical air mass over the Sahara in Africa shifts northward over the Mediterranean region and blocks maritime air masses and cyclonic systems. This shifting of stable, warm-to-hot, dry air over an area in summer and away from these regions in the winter creates a pronounced dry-summer and wet-winter pattern. The *Mediterranean* climates are found where at least 70% of annual precipitation occurs during the winter months. Cool offshore currents (the California Current, Canary Current, Peru Current, Benguela Current, and West Australian Current) produce stability in overlying maritime tropical air masses along west coasts, poleward of subtropical high pressure. The world climate map shows *Mediterranean* climates along the western continental margins.

The productivity within this climatic type is exemplified by California, which is the number one producer of 42 different commodities in the United States, and for at least 20 of these, California production is 99 percent of the total for the entire nation. The need for irrigation was perceived early, with the construction of canals in the Central Valley of California in the 1850s. Those *Mediterranean* climates that are bounded by mountains have an additional advantage in that the mountains capture winter snow pack, which melts gradually throughout the spring, releasing needed waters in the dry months of summer. The construction of reservoirs to delay runoff and capture winter precipitation is common in these regions.

The need for enormous expenditures of capital and technology is obvious.

15. Which climates are characteristic of the Asian monsoon region?

Tropical monsoon climates and humid subtropical winter-dry climates are related to the winter-dry, seasonal pulse of the monsoons and extend poleward from the tropical savanna climates. The wettest summer month as receiving 10 times more precipitation than the driest winter month. Cherrapunji, India, mentioned in Chapter 5 as receiving the most precipitation in a single year, is an extreme example of a *tropical monsoon* climate. In that location the contrast between the dry and wet monsoons is most severe, ranging from dry winds in the winter to torrential rains and floods in the summer. Downstream from the Assam Hills, such heavy rains produced floods in Bangladesh in 1988 and 1991, among other years. A representative station of the *humid subtropical winter-dry and wet summer* climate regime is Ch'engtu, China. Figure 6.11 demonstrates the strong correlation between precipitation and the high Sun of summer.

16. Explain how a *marine west coast* climate type can occur in the Appalachian region of the eastern United States.

An interesting anomaly relative to the *marine west coast* climate occurs in the eastern United States. Increased elevation in portions of the Appalachian highlands moderates summer temperatures in the *humid continental* classification, producing a *marine west coast* designation. The climograph for Bluefield, West Virginia, (Figure 6.13a and b), shows that its temperature and precipitation patterns are a marine west coast type, despite its location in the east. Vegetation similarities between the Appalachians and the Pacific Northwest are quite noticeable, enticing many emigrants who relocate from the East to settle in this climatically familiar environment in the west.

17. What role do offshore currents play in the distribution of the *marine west coast* climates? What type of fog is formed in these regions?

The dominant air mass in the *marine west coast* climates is the maritime polar, which is cool, moist, and unstable. Figure 6.5 shows the presence of cool offshore ocean currents. The weather systems that form along the polar front reach these regions throughout the year, making weather quite unpredictable. Coastal fog, totaling a month or two of days per year, is a part of the moderating marine influence. Frosts are possible and tend to shorten the growing season. These are advection fogs as discussed in Chapter 5 in the section on clouds and fog.

18. Discuss the climatic conditions for the coldest places on Earth outside the poles.

The *microthermal subarctic* climates occur only within Russia. The average temperature for the coldest month is lower than –38°C (–36.4°F). A typical *microthermal subarctic* station is Verkhoyansk, Siberia (Figure 6.18; Figure 3.24). For four months of the year average temperatures fall below –34°C (–29.2°F). Verkhoyansk frequently reaches minimum winter temperatures that are lower than –68°C (–90°F). However, as pointed out in Chapter 3, higher summer temperatures in the same area produce the world's greatest annual temperature range from winter to summer, a remarkable 63C° (113.4F°).

The *polar marine* climates are described in Shear, James A., "The Polar Marine Climate," *Annals of the Association of American Geographers* Vol. 54, No. 3 (September 1964). As with all classifications, they are portrayed as a compromise between detail and simplicity. Shear states that "It is a climate so cold that snow may fall on any day of the year, although warm enough for the mean temperature of no month to be below freezing."

_____

• *Explain* the precipitation and moisture efficiency criteria used to determine the arid and semiarid climates and *locate* them on a world map.

19. In general terms, what are the differences among the four desert classifications? How are moisture and temperature distributions used to differentiate these subtypes?

See the temperature criteria and moisture distribution inset graphs in Figure 6.21, p. 210. The major subdivisions are the *arid deserts*, where PRECIP is less than one-half of POTET, and the *semiarid steppes*, where PRECIP is more than one-half of POTET. To better approximate the dry climates, Köppen developed simple formulas to determine the usefulness of rainfall based on the season in which it falls—whether it falls principally in the summer with a dry winter, or in the winter with a dry summer, or whether the rainfall is evenly distributed. Winter rains are the most effective because they fall at a time of lower POTET.

20. Relative to the distribution of dry climates, describe at least three locations where they occur across the globe and the reasons for their presence in these locations.

The *dry desert and semiarid steppe* climates occupy more than 35% of Earth's land area, clearly the most extensive climate over land. The world climate map reveals the pattern of Earth's dry climates, which

cover broad regions between 15° and 30° N and S. In these areas the subtropical high-pressure cells predominate, with subsiding, stable air and low relative humidity. Under generally cloudless skies, these subtropical deserts extend to western continental margins, where cool, stabilizing ocean currents operate offshore and summer advection fog forms.

_____

• *Outline* future climate patterns from forecasts presented and *explain* the causes and potential consequences.

21. Explain climate forecasts. How do general circulation models (GCMs) produce such forecasts?

The 2001 IPCC *Third Assessment Report* predicts a range of average warming between 1990 to 2100 from a "low forecast" of 1.4 C° (2.5 F°), to a "high forecast" of 5.8 C° (10.4 F°). The middle case of 3.6 C° (6.5 F°) represents a significant increase in global land and ocean temperatures and will produce significant consequences.

A product of the 1992 Earth Summit in Rio de Janeiro, the largest environmental gathering of countries ever, was the United Nations Framework Convention on Climate Convention (FCCC). The leading body of the Convention is the Conference of the Parties (COP) operated by the countries that ratified the FCCC, some 170 by 1998. Subsequent meetings were held in Berlin (COP-1, 1995) and Geneva (COP-2, 1996). These meetings set the stage for COP-3 in Kyoto, Japan, December 1997, where 10,000 participants adopted the Kyoto Protocol by consensus. The latest gatherings to refine agreements were *COP-6* held in the Hague in late 2000 and *COP-7* in Marrakech, Morocco, in 2001, and *COP-8* in New Delhi in 2002. Seventeen national academies of science endorsed the Kyoto Protocol. (For updates on the status of the Kyoto Protocol see **http://www.unfccc.int/ resource/kpstats.pdf.**)

The goal, simply and boldly stated is to "...prevent dangerous anthropogenic interference with the climate system." In 2001 and 2002 the U.S. Administration abandoned the protocol process and the agreement reached at COP-7. However, a 2001 National Research Council report affirmed conclusions about human-induced climate change, saying that the IPCC Third Assessment Report "...accurately reflects the current thinking of the scientific community on this issue".

This momentum lead to Earth Summit 2002 (**http://www.earthsummit2002.org/**) in Johannesburg, South Africa, with an agenda including climate change, freshwater, gender issues, global public goods, HIV/AIDS, sustainable finance, and the five Rio Conventions.

If carbon dioxide concentrations reach 550 ppm by the year 2020 (they are presently nearing 370 ppm), a temperature increase of 3 C° (5.4 F°) is forecast for the equatorial regions. That equatorial increase translates to a high-latitude increase of about 10 C° (18 F°). Remarkably, Canada was hit with their warmest summer and year, and second warmest winter, in history in 1998—average temperatures in some areas were 2.5 C° above normal, 1.0 C° overall! This follows model predictions for high-latitude warming.

As the standard scientific reference on climate change the IPCC uses the following words to indicate levels of confidence: virtually certain (greater than a 99% chance the result is true), very likely (90–99% chance), likely (66–90% chance), medium likelihood (33–66% chance), unlikely (10–33% chance), and very unlikely (1–10% chance).

Notes on Climate Change. The reason for inclusion of this section in a physical geography text is obvious, for we live in an era of change in the physical environment. Change has occurred throughout Earth history. The critical difference now is the rate of change is faster than at any time scientists have studied. Understanding this rapid pace of human-forced change is critical if we are going to be able to formulate a response plan or a course of action for adaptation, and, if possible, to implement measures to mitigate some degree of these unwanted changes.

Scientists looking back over the past 1000 years find the rate of temperature change over the last 100 years to be greatly accelerated. One question is whether the temperature changes that are occurring are part of a natural pattern or one cause by human activity, namely increased greenhouse gases in the atmosphere. Among many sources check this recent study: R. J. Stouffer, S. Manabe, and K. Ya. Vinnikov, "Model Assessment of the Role of Natural Variability in Recent Global Warming," *Nature*, 367, February 17, 1994: 634–36. They state: "Assuming that the model is realistic, these results suggest that the observed trend is not a natural feature of the interaction between the atmosphere and oceans." This places the focus on human-induced forcing of temperatures. It is the sustained rate that is remarkable. A recent report demonstrated that atmospheric greenhouse gas concentrations and aerosol loading are responsible for the warming of the past 100 years and that they are not some aspect of a natural interaction of the atmosphere-ocean system.

Intergovernmental Panel on Climate Change (IPCC) reports are available along with many other supporting documents. A new update assessment is to be published in 2007. These are essential volumes for any research library on this topic. They form the foundation for the scientific consensus that now exists on human-forced climate change. (All of these are available in hardback as well; these ISBNs are for softcover):

1. Houghton, J.T., Jenkins, G.J. and J.J. Ephraums, editors. *Climate Change—The IPCC Scientific Assessment*, Working Group I, World Meteorological Organization/United Nations Environment Programme. Port Chester, N.Y.: Cambridge University Press, 1991. (Paper-ISBN: 0-521-40720-6; 1-800-431-1580.)
2. Intergovernmental Panel on Climate Change. *Climate Change—The IPCC Response Strategies*, Working Group III. World Meteorological Organization/United Nations Environment Programme. Covelo, CA: Island Press, 1991. (Paper-ISBN: 1-55963-102-3; 1-800-828-1302.)
3. Tegart, W.J. McG., Sheldon, G.W. and D.C. Griffiths. *Climate Change—The IPCC Impacts Assessment,* Working Group II. World Meteorological Organization/United Nations Environment Programme. Portland, OR: International Specialized Book Service/Australian Government, 1991. (Paper-ISBN: 0-644-13496-6; 1-503-286-3093.)
4. Houghton, J.T., Jenkins, G.J. and J.J. Ephraums, editors. *Climate Change 1992—The Supplementary Report to the IPCC Scientific Assessment. World Meteorological Organization/United Nations Environment Programme.* New York: Cambridge University Press, 1992. (Paper-ISBN: 0-521-43829-2; 1-800-431-1580.) This report contains many color GCM maps.
5. Houghton, J.T., Filho, L.G.M., Callander, B.A., Harris, N., Kattenberg, A., Maskell, K., editors. *Climate Change 1995—The Science of Climate Change*, contributions of Working Group I to the Second Assessment Report of the IPCC. New York: Cambridge University Press, 1995. (Paper-ISBN: 0-521-56436-0; 1-800-431-1580.)
6. Watson, R.T., Zinyowera, M.C., Moss, R.H., Dokken, D.J., editors. *Climate Change 1995—Impacts, Adaptations, and Mitigation of Climate Change: Scientific-Technical Analyses*, contribution of Working Group II to the Second Assessment Report of the IPCC. New York: Cambridge University Press, 1995. (Paper-ISBN: 0-521-56436-9; 1-800-431-1580.)
7. Bruce, J.P., Lee, H., and Haites, E.F., editors. *Climate Change 1995—Economic and Social Dimensions of Climate Change*, contribution of Working Group III to the Second Assessment Report of the IPCC. New York: Cambridge

University Press, 1995. (Paper-ISBN: 0-521-56854-4; 1-800-431-1580.)

8. IPCC Working Group I, *Climate Change 2001: The Scientific Basis*. Washington: Cambridge University Press, 2001.

9. IPCC Working Group II, *Climate Change 2001: Impacts, Adaptation, and Vulnerability*. Washington: Cambridge University Press, 2001.

10. IPCC Working Group III, *Climate Change 2001: Mitigation*. Washington: Cambridge University Press, 2001.

There is a wealth of web sites devoted to climate change as well. In addition to the sites listed on the text's companion site I would recommend the following:

The IPCC's main page: **http://www.ipcc.ch**

The Union of Concerned Scientists: **http://www.ucsusa.org/global_warming/**

The Arctic Climate Impact Assessment site: **http://www.acia.uaf.edu/**

The Pew Center on Global Climate Change: **http://www.pewclimate.org/**

Stephen H. Schneider in his book *Global Warming: Are We Entering the Greenhouse Century* (New York: Random House, 1989) states, "We are insulting the atmospheric environment faster than we are comprehending the effects of those insults."

A 1997 book by Ross Gelbspan, *The Heat Is On* (New York: Addison-Wesley Publishing Co., 1997, ISBN: 0-201-13295-8) overviews the entire subject of global climate change. It is not original research but a reporting of primary and original sources. The book is valuable for it discloses the funding that supports the few skeptics that get so much media attention yet have not published in peer review journals. Hundreds of thousands of dollars from big oil, big coal, mining interests, and automobile and trucking interests have poured into funding these skeptics. Good insight is presented into the Marshall and Cato Institutes and their reports, which are heavily referenced by the "skeptics." He followed this up in 2004 with *Boiling Point: How Politicians, Big Oil and Coal, Journalists and Activists Are Fueling the Climate Crisis—And What We Can Do to Avert Disaster* (Basic Books, 2004, 046502761X). This volume is pretty much what you would expect from the title—an introduction to the political aspects of climate change science. Please consider reading these books for perspective.

A few basic issues: First, this subject is heavily politicized, especially by those interests who think that they have the most to lose by any changes from traditional, centralized, energy-production patterns (big oil, big coal, and transportation). Much disinformation is spread by popular, opinion-rich media shows on radio and television. Since 1994, Congress has been dominated by representatives willing to elevate unpublished and unfounded opinion from several prominent "skeptics" over accepted scientific consensus. Congressmen Rohrabacher, Doolittle, and Delay conducted hearings that would be laughable had they not led to proposed legislation. They reject all of the IPCC work, have promoted false and misleading opinion, and even proposed terminating the United States participation in the ozone treaty—something industries such as DuPont do not favor. This is all documented in Ross Gelbspan's book *The Heat Is On* referenced earlier.

Unfortunately, most traditional energy production is based on fuels that are linked to the production of radiatively active gases, especially carbon dioxide. In Chapter 4 of this resource guide, two lists of centralized and decentralized energy source characteristics were presented. The centralized list is the one being promoted for the world by the transnational corporations.

Second, we are asked to treat radiatively active gases, such as CO_2, as if they were somehow an arrestee with full civil rights to a trial and a presumption of innocence until guilt is proven; or, a presumption of open-ended risk assessment. In an ideal world the CO_2 would be assumed guilty until its supporters proved it was innocent. With such serious issues, we might want to ask why our representative political system places most risk on the side of *error* rather than on the side of *caution*. The burden of proof is placed on the public (the potential victim) to prove that something is disruptive to a sustainable environment, and not on those entities that produce the contaminant or force unwanted change.

An objective observer might wonder if modern society's unconscious proposition seems simply to be, remodel the ecological texture of North America and the planet and then proceed with an experiment in adaptation! Before we embark on such a one-time only experiment, perhaps the perpetrators need to prove a higher level of sustainability for their actions. Admittedly, this requires a long-term view, such as the one you sense when you look into a child's eyes.

22. Describe the potential climatic effects of global warming on polar and high-latitude regions. What are the implications of these climatic changes for persons living at lower latitudes?

Perhaps the most pervasive climatic effect of increased warming would be the rapid escalation of ice melt. The additional water, especially from continental ice masses that are grounded, would raise sea level worldwide. Scientists are currently studying the ice sheets of Greenland and Antarctica for possible changes in the operation of the hydrologic cycle, including snowlines and the rate at which icebergs break off (calve) into the sea. The key area being watched is the

West Antarctic ice sheet, where the Ross Ice Shelf holds back vast grounded ice masses.

An announcement that 43% of the Arctic Ocean ice had melted back since 1970 shocked researchers and the public. Significant change, occurring at unprecedented rates, appears to be underway at higher latitudes—as forecasted by GCMs.

Loss of polar ice mass, augmented by melting of alpine and mountain glaciers, will affect sea-level rise. A quick survey of world coastlines shows that even a moderate rise could bring change of unparalleled proportions. At stake are the river deltas, lowland coastal farming valleys, and low-lying mainland areas, all contending with high water, high tides, and higher storm surges. There will be both internal and international migration of affected populations, spread over decades, away from coastal flooding if sea levels continue to rise.

A study of Alaskan glaciers published in 2002 looked at the rate of glacier-wide thickness loss for 67 glaciers between the mid-1950s and 1995, and for 28 glaciers between 1995 and 2001. A total volume change was estimated to be -52 ± 15 km^3 per year (-12.5 ± 3.6 mi^3), equaling a water equivalent of -96 ± 35 km^3 per year (-23 ± 8.4 mi^3). These losses alone account for a rise in global sea level of 0.14 ± 0.04 mm and 0.27 ± 0.10 mm per year (0.55 and 1.06 in. per year). This is twice the volume estimated to be coming from the entire Greenland ice sheet during the 1995 to 2001 time period. This means that about 9% of global sea level rise is coming from the Alaskan meltdown.

23. How is climatic change affecting agricultural? Natural environments? Forests? The possible spread of disease?

Modern single-crop agriculture is more delicate and susceptible to temperature change, water demand and irrigation needs, and soil chemistry than is traditional multicrop agriculture. Specifically, the southern and central grain-producing areas of North America are forecast to experience hot and dry weather by the middle of the next century as a result of higher temperatures. An increased probability of extreme heat waves is forecast for these U.S. grain regions. Also, available soil moisture is projected to be at least 10% less throughout the midlatitudes over the next 30 years than present levels. Scientists are considering changing to late-maturing, heat-resistant crop varieties and adjusting fertilizer applications and irrigation.

Biosphere models predict that a global average of 30% of the present forest cover will undergo major species redistribution, the greatest change occurring in high latitudes. Many plant species are already "on the move" to more favorable locations. Land dwellers must also adapt to changing forage. Warming is already stressing some embryos as they reach their thermal limit. Particularly affected are amphibians, whose embryos develop in shallow water. The warming of large bodies of water may benefit some species, and harm others.

Recent studies suggest that climate change may affect health on a global basis. Populations previously unaffected by malaria, schistosomiasis, sleeping sickness, and yellow fever would be at greater risk in subtropical and midlatitude areas.

24. What are the present actions being taken to delay the effects of global climate change? What is the Kyoto Protocol?

A product of the 1992 Earth Summit in Rio de Janeiro, the largest environmental gathering of countries ever, was the United Nations Framework Convention on Climate Convention (FCCC). The leading body of the Convention is the *Conference of the Parties* (COP) operated by the countries that ratified the FCCC, some 170 by 1998. Subsequent meetings were held in Berlin (*COP-1*, 1995) and Geneva (*COP-2*, 1996). These meetings set the stage for *COP-3* in Kyoto, Japan, December 1997, where 10,000 participants adopted the *Kyoto Protocol* by consensus. Seventeen national academies of science endorsed the Kyoto Protocol. The latest gatherings to refine agreements were *COP-6* held in the Hague in late 2000, *COP-7* in Marrakech, Morocco, in 2001, *COP-8* in New Delhi in 2002, *COP-9* in Milan in 2003, followed by *COP-10* in Buenos Aires 2004, and *COP-11* in Montreal in December 2005.

The Kyoto Protocol binds more-developed countries to a collective 5.2% reduction in greenhouse gas emissions as measured at 1990 levels for the period 2008 to 2012. Within this group goal, various countries promised cuts: Canada is to cut 6%, the European Union 8%, and Australia 8%, among many others. The *Group of 77* countries plus China favor a 15% reduction by 2010. With Russian ratification in November 2004, the Kyoto Protocol is now international law with 156 country signatories; this is without United States participation or support. (For updates on the Kyoto Protocol, see **http://unfccc.int/2860.php**.)

The U.S. President asked the National Research Council to assess the level of IPCC science. The NRC quickly responded in a report affirming conclusions about human-induced climate change, saying that the IPCC Third Assessment Report "...accurately reflects the current thinking of the scientific community on this issue..." (NRC, *Climate Change Science, An Analysis of Some Key Questions*, Washington: National Academy Press, May 2001).

The Protocol is far-reaching in scope, including calling for international cooperation in meeting goals, technology development and transfers, leniency for less developed countries, "clean development" initiatives, and "emissions trading" schemes between industrialized countries and individual industries. The goal, simply and boldly stated is to "…prevent dangerous anthropogenic interference with the climate system."

_____

Overhead Transparencies

126.	-----	A **blank** climograph
127.	CO6	*Terra* satellite image Taklimakan Desert
128.	6.1	A schematic of Earth's climate system
129.	F.S. 6.1.1 a – e	Normal and El Niño patterns in the Pacific, with 5 satellite images
130.	6.2	Worldwide average annual precipitation
131.	6.3	Climate relationships
132.	6.4	Earth's climates generalized
133.	6.5	World climate classifications
134.	Tropical, Mesothermal, Microthermal-climate boxes	Tropical, Mesothermal, and Microthermal climate
135.	Polar-, and Desert-climate boxes	Polar and Desert climate
136.	6.6a, b	Locator map, climograph, and photo of Uaupés, Brazil
137.	6.7a, b	Locator map, climograph, and photo of Yangon, Myanmar
138.	6.8a, b	Locator map, climograph, and photo of Arusha, Tanzania
139.	6.9a, b	Locator map and climograph for Columbia, South Carolina
140.	6.11a, b	Locator map and climograph for Vancouver, British Columbia
141.	6.13a, b	Locator map and climograph for Bluefield, West Virginia
142.	6.14a, b	Locator map and climographs for San Francisco and Sevilla
143.	6.15a, b	Locator map and climographs for New York City and Dalian, China
144.	6.17a	Locator map and climograph for Churchill, Manitoba
145.	6.18a	Locator map and climograph for Verkhoyansk, Siberia, Russia
146.	6.21 and 6.22	Locator maps and climographs for Ar Riyad, Saudi Arabia and Albuquerque, New Mexico
147.	6.25 a, b	Global temperature trends, 1880–2002(top); temperature anomalies 1998 from 1951–1980 base line (bottom)
148.	6.26	Origins of excessive CO_2
149.	6.27	General circulation model scheme

PART THREE:
Earth's Changing Landscape Systems

Overview

Earth is a dynamic planet whose surface is actively shaped by physical agents of change. Part Three is organized around two broad systems of these agents—the internal (endogenic) and external (exogenic). The endogenic system (Chapters 8 and 9) encompasses internal processes that produce flows of heat and material from deep below the crust, powered by radioactive decay. This is the solid realm of Earth. "The Ocean Floor" chapter-opening illustration that begins Chapter 9 is used as a bridge between these two endogenic chapters. The exogenic system (Chapters 10–14) includes external processes that set water, air, waves, and ice into motion, powered by solar energy. This is the fluid realm of Earth's environment. These media are sculpting agents that carve, shape, and reduce the landscape. The content is organized along the flow of energy and material or in a manner consistent with the flow of events.

To assist in preparing lecture materials for Part 3, I recommend that you obtain a copy of *Geomorphology from Space—A Global Overview of Regional Landforms*, edited by Nicholas M. Short and Robert W. Blair, Jr. Washington, D.C.: Scientific and Technical Information Branch, National Aeronautics and Space Administration, 1986. This is available from the Superintendent of Documents, U.S. GPO. The 700-page volume has chapters on regional landform analysis associated with tectonic, volcanic, fluvial, deltaic, coastal, karst, lacustrine, eolian, glacial, and planetary landform processes. Also included are sections on geomorphology mapping and a future look at global geomorphology. This book contains thousands of images and photographs, many in full color. Each chapter is accompanied by an informative text, references, and detailed source information for the images used. It is a great resource for teaching, lecture preparation, and classroom media development. Also there are two other basic reference works—Bates, Robert L., and Julia A. Jackson eds., *Glossary of Geology*, 3rd ed., Alexandria, VA: American Geological Institute (AGI), 1987; and a topical source book: Smith, David G. ed., *The Cambridge Encyclopedia of Earth Sciences*, New York (London): Cambridge University Press, 1981.

The Dynamic Planet

The 20th century was a time of great discovery about Earth's internal structure and dynamic crust, yet much remains undiscovered. It was a time of revolution in our understanding of how the present arrangement of continents and oceans evolved. One task of physical geography is to explain the spatial implications of all this new knowledge and its effect on Earth's surface and society.

The fact that the continents are actively adrift is still astounding and perhaps unknown to many. Physical geographers have a responsibility to turn people on to this information about our dynamic planet.

Outline Headings and Key Terms

The first-, second-, and third-order headings that divide Chapter 8 serve as an outline for your notes and studies. The key terms and concepts that appear boldface in the text are listed here under their appropriate heading in bold italics. All these highlighted terms appear in the text glossary. Note the check-off box (❑) so you can mark your progress as you master each concept. Your students have this same outline in their *Student Study Guide*. The ✪ icon indicates that there is an accompanying animation or other resource on the CD.

The outline headings for Chapter 8:

✪ **Earth's Varied Landscapes: Aerial Photo gallery**
   - ❑ *endogenic system*
   - ❑ *exogenic system*

The Pace of Change

✪ **Applying Relative Dating Principles**
   - ❑ *geologic time scale*
   - ❑ *uniformitarianism*

Earth's Structure and Internal Energy

   Earth in Cross-Section
   - ❑ *seismic waves*

   Earth's Core
   - ❑ *core*

   Earth's Magnetism
   - ❑ *geomagnetic reversal*

   Earth's Mantle
   - ❑ *mantle*
   - ❑ *asthenosphere*

   Lithosphere and Crust
   - ❑ *crust*
   - ❑ *Mohorovicic discontinuity (Moho)*
   - ❑ *granite*
   - ❑ *basalt*
   - ❑ *isostasy*

The Geologic Cycle

✪ **The Rock Cycle Notebook**

✪ **Igneous Rocks Table**

✪ **Formation of intrusive igneous features**

✪ **Foliation (metamorphic rock)**
   - ❑ *geologic cycle*

   Rock Cycle
   - ❑ *mineral*
   - ❑ *rock*

   Igneous Processes
   - ❑ *igneous rock*
   - ❑ *magma*
   - ❑ *lava*
   - ❑ *pluton*
   - ❑ *batholith*

   Sedimentary Processes
   - ❑ *stratigraphy*
   - ❑ *sedimentary rocks*
   - ❑ *lithification*
   - ❑ *limestone*
   - ❑ *evaporites*

   Metamorphic Processes
   - ❑ *metamorphic rock*

Plate Tectonics

✪ **Seafloor Spreading, Subduction**

✪ **Pangaea Breakup, Plate movements**

✪ **Plate Boundaries Notebook**

✪ **India Collision with Asia**

✪ **Transform Faults, Plate Margins**

   A Brief History
   - ❑ *Pangaea*
   - ❑ *Plate tectonics*

   Seafloor Spreading and Production of New Crust

✪ **Convection in a Lava Lamp**
   - ❑ *sea-floor spreading*
   - ❑ *mid-ocean ridges*

   Subduction of Crust
   - ❑ *subduction zone*
   - ❑ *oceanic trenches*

   The Formation and Breakup of Pangaea

✪ **Plate Motions Through Time**

   Plate Boundaries

✪ **Motion at Plate Boundaries**

✪ **Correlating Processes and Plate Boundaries**

✪ **Forming a Divergent Boundary**

   Transform faults, plate margins
   Earthquakes and Volcanoes
   Hot Spots

✪ **Hot Spot Volcano Tracks**
   - ❑ *hot spots*
   - ❑ *geothermal energy*

Summary and Review

News Reports

The URLs related to this chapter of *Elemental Geosystems* can be found at
http://www.prenticehall.com/christopherson

Key Learning Concepts

After reading the chapter and using the Student Study Guide, the student should be able to:

• *Distinguish* between the endogenic and exogenic systems, *determine* the driving force for each, and *explain* the pace at which these systems operate.

• *Diagram* Earth's interior in cross section and *describe* each distinct layer.
• *Illustrate* the geologic cycle and *relate* the rock cycle and rock types to endogenic and exogenic processes.
• *Describe* Pangaea and its breakup and *relate* several physical proofs that crustal drifting is continuing today.
• *Portray* the pattern of Earth's major plates and *relate* this pattern to the occurrence of earthquakes, volcanic activity, and hot spots.

Annotated Chapter Review Questions

• *Distinguish* **between the endogenic and exogenic systems,** *determine* **the driving force for each, and** *explain* **the pace at which these systems operate.**

1. To what extent is Earth's crust actively building at this time in its history?

The U.S. Geological Survey reports that, in an average year, continental margins and seafloors expand by 1.9 km^3 (0.46 mi^3). But, at the same time, 1.1 km^3 (0.26 mi^3) are consumed, resulting in a net addition of 0.8 km^3 (0.2 mi^3) to Earth's crust. The results are irregular patterns of surface fractures, the occurrence of earthquakes and volcanic activity, and the formation of mountain ranges.

2. Define the endogenic and the exogenic systems. Describe the driving forces that energize these systems.

The endogenic system (Chapters 8 and 9) encompasses internal processes that produce flows of heat and material from deep below the crust, powered by radioactive decay. This is the solid realm of Earth. The exogenic system (Chapters 10–14) includes external processes that set air, water, and ice into motion, powered by solar energy. This is the fluid realm of Earth's environment. These media are sculpting agents that carve, shape, and reduce the landscape—all under the pervasive influence of gravity.

3. How is the geologic time scale organized? What is the basis for the time scale in relative and absolute terms? What era, period, and epoch are we living in today?

The geologic time scale (Figure 8.1) reflects currently accepted names and the relative and absolute time intervals that encompass Earth's history (eons, eras, periods, and epochs). The sequence in this scale is based upon the relative positions of rock strata above or below one another. An important general principle is that of superposition, which states that rock and sediment always are arranged with the youngest beds "superposed" near the top of a rock formation and the oldest at the base—if they have not been disturbed. The

absolute ages on the scale, determined by scientific methods such as dating by radioactive isotopes, are also used to refine the time-scale sequence. The figure presents important events in Earth's life history along with the geologic time scale.

For interesting historical background see Lawrence Badash's "The Age-of-the-Earth Debate," *Scientific American* (August 1989): pp. 90–96. Also see a good succinct review in John Thackray's *The Age of the Earth*, London: Her Majesty's Stationery Office for the Institute of Geological Sciences, 1980 (ISBN 0-11-884077-0), available in the U.S. from Cambridge University Press, New York. Although it is filled with European terminology, a good treatment of chronostratic and chronometric time scales and related dating methods is in Harland, W. Brian, Armstrong, Richard L., *et al.*, *A Geologic Time Scale*, New York (London): Cambridge University Press (1990).

4. Describe uniformitarianism in Earth's development. How can this flow of events and time be interrupted?

The guiding principle of Earth science is uniformitarianism, first proposed by James Hutton in his *Theory of the Earth* (1795) and later amplified by Charles Lyell in his *Principles of Geology* (1830). Uniformitarianism assumes that *the same physical processes active in the environment today have been operating throughout geologic time.* "The present is the key to the past" describes this principle. Evidence unfolding from modern scientific exploration and from the landscape record of volcanic eruptions, earthquakes, and Earth's processes support uniformitarianism.

However, geologic time is punctuated by dramatic events, such as massive landslides, earthquakes, volcanic episodes, and extraterrestrial asteroid impacts. Within the principle of uniformitarianism, these localized catastrophic events occur as small interruptions in the generally uniform processes that shape the slowly evolving landscape. Here, the *punctuated equilibrium* (interruptions in the flow of events; jumps to new system operation levels) concept studied in the life sciences and paleontology might apply to aspects of Earth's long developmental history.

• *Diagram* **Earth's interior in cross section and** *describe* **each distinct layer.**

5. Make a simple sketch of Earth's interior, label each layer, and list the physical characteristics, temperature, composition, and range of size of each on your drawing.

See details in Figures 8.2, 8.3, and 8.4 as a basis for this sketch. For an up-to-date survey of the science of Earth's structure, see "Research: From the

Core to the Crust," in *Geotimes*, July 1999 issue (Vol. 44, No. 7).

6. What is the present thinking on how Earth generates its magnetic field? Is this field constant, or does it change? Explain the implications of your answer.

The fluid outer core generates at least 90% of Earth's magnetic field and the magnetosphere that surrounds and protects Earth from the solar wind. A present hypothesis by scientists from Cambridge University details spiraling circulation patterns in the outer core region that are influenced by Earth's rotation; this circulation generates electric currents, which in turn induce the magnetic field. An intriguing feature of Earth's magnetic field is that it sometimes fades to zero and then returns to full strength with north and south magnetic poles reversed! In the process, the field does not blink on and off but instead oscillates slowly to nothing and then slowly regains its strength. (New evidence suggests the field fades slowly to zero, then when it returns it tends to do so abruptly.) This magnetic reversal has taken place nine times during the past 4 million years and hundreds of times over Earth's history. The average period of a magnetic reversal is 500,000 years, with occurrences as short as several thousand years possible.

Earth's magnetic field presently is losing strength at the rate of approximately 7% per 100 years. The field was about 40% stronger 2000 years ago according to the latest published research.

7. Describe the asthenosphere. Why is it also known as the plastic layer? What are the consequences of its convection currents?

The extreme upper mantle, just below the crust, is known as the asthenosphere, or plastic layer. It contains pockets of increased heat from radioactive decay and is susceptible to convective currents in these hotter (and therefore less dense) materials. The depths affected by these convection currents are the subject of much scientific speculation. Because of this dynamic condition, the asthenosphere is the least rigid region of the mantle, with densities averaging 3.3 g/cm^3. This section of the mantle is known as the plastic layer due to its dynamic activity. About 10% of the asthenosphere is molten in asymmetrical patterns and hot spots. Think of Earth's outer crust (densities of 2.7 g/cm^3 for continental crust and 3.0 g/cm^3 for oceanic crust) as floating on the denser layers beneath, much as a boat floats on water. With a greater load (e.g., ice, sediment, mountains), the crust tends to ride lower in the asthenosphere. Convection currents in the asthenosphere disturb the overlying crust and create tectonic activity. In return, the movement of the crust-collision, divergence, may influence currents in the mantle.

8. What is a discontinuity? Describe the principal discontinuities within Earth.

A discontinuity is a place where a change in physical properties occurs between two regions deep in Earth's interior. A transition zone of several hundred kilometers marks the top of the outer core and the beginning of the mantle. Scientists at the California Institute of Technology analyzed the behavior of more than 25,000 earthquakes and determined that this transition area is bumpy and uneven, with ragged peak-and-valley-like formations. Some of the motions in the mantle may be created by this rough texture at what is called the Gutenberg discontinuity. The boundary between the crust and the rest of the lithospheric upper mantle is another discontinuity called the Mohorovicic discontinuity, or Moho for short, named for the Yugoslavian seismologist who determined that seismic waves change at this depth, owing to sharp contrasts of materials and densities.

9. Define isostasy and isostatic rebound, and explain the crustal equilibrium (balance between buoyancy and gravity) concept.

The principle of buoyancy (that something less dense, like wood, floats in denser things like water) and the principle of balance were further developed in the 1800s into the important principle of isostasy to explain certain movements of Earth's crust. The entire crust is in a constant state of compensating adjustment, or isostasy, slowly rising and sinking in response to its own weight, and pushed and dragged about by currents in the asthenosphere (Figure 8.4).

10. Diagram the uppermost mantle and crust. Label the density of the layers in gm/cm^3. What two types of crust were described in the text in terms of rock composition.

See Figures 8.2c and 8.4 as the basis for this diagram. The two types of crust discussed in the text were oceanic crust, composed of basalt, a rock high in silica and magnesium (earning its name as *simatic* crust), and continental crust, composed mostly of granite, a rock high in silica and aluminum (earning its name as *sialtic* crust).

• *Illustrate* the geologic cycle and *relate* the rock cycle and rock types to endogenic and exogenic processes.

11. Illustrate the geologic cycle and define each component: rock cycle, tectonic cycle, and hydrologic cycle.

See Figure 8.5 as the basis of this illustration.

12. What is a mineral? A mineral family? Name the most common minerals on Earth. What is a rock?

A mineral is an element or combination of elements that forms an inorganic natural compound. A mineral can be described with a specific symbol or formula and possesses specific qualities. Silicon (Si) readily combines with other elements to produce the silicate mineral family, which includes quartz, feldspar, amphibole, and clay minerals, among others. Another important mineral family is the carbonate group, which features carbon in combination with oxygen and other elements such as calcium, magnesium, and potassium. Of the nearly 3000 minerals, only 20 are common, with just 10 of those making up 90% of the minerals in the crust. A rock is an assemblage of minerals bound together (such as granite, containing silica, aluminum, potassium, calcium, and sodium) or sometimes a mass of a single mineral, such as rock salt.

Some background notes on economic aspects of rocks and minerals. Many different types of rock are useful and of economic importance to society. Clays of certain grades are used for pipe-making and pottery, some very fine clays are used to coat paper, such as the paper on which the textbook is printed. Pure sands, high in quartz content, are processed in glass making. Sands and gravels are an important aggregate in cement and for building construction. Lime derived from limestone is used in the making of cement and in agriculture. Phosphates from marine shales and limestone are important in the making of fertilizer. Gypsum, an evaporite derived from deposits related to sea water, is used in plaster. Marble, granite, and limestone are used as construction material, with pure white marble preferred for sculpture and decorative building facades. And, of course, salt (NaCl) is of importance to civilization throughout history. The rock cycle is intricately woven into society in many ways.

An **ore** is a body of rock which contains minerals sought by society. If the concentration of the desired mineral is high enough, mining becomes economically feasible. A major concentration process in nature for iron, lead, zinc, mercury, copper, and other minerals involves hydrothermal solutions. Sometimes associated with an intrusive igneous pluton, and sometimes with contact metamorphism, high temperature moisture solutions dissolve minerals and carry them through cracks, joints, and fractures. These valuable elements precipitate out in vein formations. In many places, individual veins of ore are traceable back to the parent igneous pluton. Sometimes these ore deposits occur in association with other vein-filling minerals such as quartz and calcite.

In other areas, *hydrothermal* solutions disseminate the mineral precipitate throughout an ore body, producing a low concentration of the desired

mineral in a large mass of ore. Large-scale mining methods are used to extract such disseminated minerals. The Bingham Open Pit Copper Mine west of Salt Lake City, Utah is a prime example. The copper ore was so low-grade that economics dictated location of the concentrator, smelter, and refinery near the mine, thus reducing transportation costs. To get 6.4 kg of copper, almost 900 kg of ore had to be processed, which required the removal of 2.04 metric tons of overburden (14 lbs. of copper, 2000 lbs. of ore, 4500 lbs of overburden). Low copper prices worldwide kept the operation, along with other western copper mines, closed during much of the 1980s.

Relative to the formation of mineral deposits on the ocean floor, current scientific thinking points directly to the actions of plate tectonics. For the first time, scientists aboard deep-sea submersibles saw mineral deposits actually forming on the ocean floor. Hot solutions of minerals spew from ocean floor vents that are associated with the mid-ocean ridge system. Minerals precipitate out when the hot solution comes into contact with the near-freezing temperatures of the deep ocean.

Plate tectonics carries these accumulated deposits toward eventual collision and subduction beneath the continents. In areas where some of the "cargo" on the plate is not subducted but is instead raised and pasted onto the continental mass, the mineral-rich content is readily visible. On the island of Cypress in the Mediterranean Sea, deposits of copper were mined for the past 4000 years by various civilizations. Cypress is composed of old seafloor pressed upward by the African-Eurasian collision.

The deposits that do subduct melt with the diving plate and work their way toward the surface, cooling and dissolving according to the specific nature of each element or mineral involved. Hot water trapped in the rock does the rest, dispersing the minerals in veins or disseminating the mineral ore bodies as we discussed above.

Other formation processes were at work in Sudbury, Ontario, Canada. Nickel, iron, and copper came up with a mafic (high in magnesium and iron), intrusive body of magma that cooled and began crystallization. These elements, along with other minerals, settled out in the magma chamber in specific layers, forming a very rich resource body and the basis for an active mining district.

13. Describe igneous process. What is the difference between intrusive and extrusive types of igneous rocks?

Rocks that solidify and crystallize from a molten state are called igneous rocks. Most rocks in the crust are igneous. They form from magma, which is molten rock beneath the surface (hence the name igneous, which means *fire-formed* in Latin). Magma is fluid, highly gaseous, and under tremendous pressure. It is either intruded into preexisting crustal rocks, known as country rock, or extruded onto the surface as lava. The cooling history of the rock—how fast it cooled, and how steadily the temperature dropped—determines its texture and degree of crystallization. These range from coarse-grained (slower cooling, with more time for larger crystals to form) to fine-grained or glassy (faster cooling).

Notes on Igneous Rocks. The categorization of igneous rocks is usually done by mineral composition and texture. The two broad categories are:

1. *Felsic igneous rocks*—derived both in composition and name from *fel*dspar and *silica* (SiO_2). Felsic minerals are generally high in silica, aluminum, potassium, and sodium, with low melting points. Rocks formed from felsic minerals generally are lighter in color and density than mafic mineral rocks.

2. Mafic igneous rocks—derived both in composition and name from *ma*gnesium and *ferric* (Latin for iron). Mafic minerals are low in silica and high in magnesium and iron, with high melting points. Rocks formed from mafic minerals are darker in color and of greater density than felsic mineral rocks.

The same magma which produces coarse-grained granite when slowly cooled beneath the surface also forms a fine-grained rhyolite as its rapidly cooled volcanic counterpart. If it cools rapidly, magma having a silica content comparable to granite and rhyolite may form the dark, smoky, glassy-textured rock called *obsidian* or volcanic glass. Another glassy rock called *pumice* forms when escaping gases bubble a frothy texture into the lava. Pumice is full of small openings, is light in weight, and low enough in density to float in water.

On the mafic side, basalt is the most common fine-grained extrusive igneous rock. It comprises the bulk of the ocean floor and appears in lava flows such as those on the Galápagos Islands. Its intrusive counterpart, formed by slow cooling of the parent magma, is *gabbro*.

14. Briefly explain how the rock formation in Figure 8.9a demonstrates through its layers a record of past climates.

This sedimentary sandstone, with siltstone below, demonstrates different iron content through its coloration change, therefore different parent materials. The weaker siltstones below are weathering faster, leaving the mass above developing into a balanced rock form. Drier times in the past were marked by sand dunes. These dune forms are lithified into rock near the upper third of the rock, layered at an angle. Wetter times, when more flowing water was present, are marked by horizontal beds, laid down as sediment and lithified into stone. The discontinuity between the horizontal and angular layers must have been a time of change as parts of the record were eliminated through weathering and erosional processes.

15. Briefly describe sedimentary processes and lithification. Describe the sources and particle sizes of sedimentary rocks.

Most sedimentary rocks are derived from preexisting rocks, or from organic materials such as bone and shell that form limestone, mud that becomes compacted into shale, and ancient plant remains that become compacted into coal. The exogenic processes of weathering and erosion generate the material sediments needed to form these rocks. Bits and pieces of former rocks—principally quartz, feldspar, and clay minerals—are eroded and then mechanically transported (by water, ice, wind, and gravity) to other sites where they are deposited. In addition, some minerals are dissolved into solution and form sedimentary deposits by precipitating from those solutions; this is an important process in the oceanic environment. The cementation, compaction, and hardening of sediments into sedimentary rocks are called lithification. Various cements fuse rock particles together; lime ($CaCO_3$, or calcium carbonate) is the most common, followed by iron oxides (Fe_2O_3) and silica (SiO_2). Particles also can unite by drying (dehydration), heating, or chemical reactions. The two primary sources of sedimentary rocks—the mechanically transported bits and pieces of former rock and the dissolved minerals in solution—are known as clastic sediments and chemical sediments, respectively.

The example of evaporites forming in Death Valley, shown in the documentary pair of photos in Figure 8.10, was a dramatic experience. Another pair of photos taken over these same two days appears in Chapter 13. During the 1982–83 El Niño event the southwest and California experienced record precipitation. On a day in 1983 Death Valley received 2.57 cm, or about 55% of the normal expected amount for an entire year. I waited overnight at a roadblock because the road into the valley was closed with playa flooding over the pavement. My wait was rewarded with incredible scenes of fluvial action in the desert.

The lake you see in Figure 8.10a is approximately 3 km wide and 8 km long and only several centimeters deep. I walked out about a kilometer. The natural sorting process that began at the top of the alluvial fans had left large rocks and grains behind, leaving only a very fine clay—as fine as face powder—to squish between my toes. The scene was strange because the blocked roads had let few into the valley, making the aloneness, quiet, and stillness almost overwhelming. I returned one month later and matched the pictures using a slide viewer in one hand and a camera in the other to produce the picture in (b). A bed of borated-salt precipitate approximately 2 cm thick replaced the vast reflective water surface. The salt surface cracked underfoot exposing those very fine clays. As a result of the two trips I have matching pairs of about 50 scenes throughout Death Valley demonstrating that water is the major erosional force in the desert, however infrequently it occurs!

Mono Lake Briefing. Another type of hydro-thermal activity produces evaporites and sedimentary formations near the Nevada-California border at **Mono Lake**, an alkaline lake high in carbonates and sulfates that presently has an overall salinity of 95‰ (the ocean average is 35‰, or 35 parts per thousand). Carbonates in Mono Lake interact with calcium-rich hot springs to produce a rock called *tufa*. The tufa deposits grew tall in Mono Lake as long as the springs were beneath the surface of the lake.

However, since the 1940s, tributary streams have been diverted out of the region, lowering the lake's surface more than 20 m (65 ft), reducing its surface area by more than half, exposing these chemical sedimentary deposits and alkaline shorelines, and damaging the lake's natural ecology and habitats for wildlife and birds, particularly birds that perished in large numbers in the early 1980s. Some tufa towers now rise over 10 m (30 ft) above the shoreline, as shown in the foreground of Figure 5.18h. Strong winds in the area blow across the newly formed alkali flats of evaporites, producing days of severe air pollution.

After much litigation and a case that went to the California Supreme Court, the state's water law was significantly changed. Century-old appropriative water rights (first one at water's edge withdraws whatever they want) were overturned in favor of placing this body of water in the *public trust*—benefits to the many overshadowed rights of a few. A master plan was completed and agreed to by all parties and the future looks bright for this water-troubled region. Tributary flows were restored by court order in April 1991, and the lake level is rising. Additional court and state agency rulings in 1994 assured victory for the lake.

By May 2, 2006, the lake level stood at 1945.4 m (6383.2 ft), an increase of more than 2.5 m since 1994. The tributaries are restoring habitats, bird populations are increasing in numbers, and lake ecology improving—a real success story! (Mono Lake Committee office, 760-647-6595, Lee Vining, CA, **http:/www.monolake.org**).

---

16. What is metamorphism and how are metamorphic rocks produced? Name some original parent rocks and metamorphic equivalents.

Any rock, either igneous or sedimentary, may be transformed into a metamorphic rock by going through profound physical and/or chemical changes under increased pressure and temperature. (The name metamorphic comes from the Greek, meaning to *change form.*) Metamorphic rocks generally are more compact than the original rock and therefore are harder and more resistant to weathering and erosion. See Table 8.1, p. 268.

• *Describe* **Pangaea and its breakup and** *relate* **several physical proofs that crustal drifting is continuing today.**

17. Briefly review the history of continental drift, sea-floor spreading, and the all-inclusive plate tectonics theory. What was Alfred Wegener's role?

In 1912, German geophysicist and meteorologist Alfred Wegener publicly presented in a lecture his idea that Earth's landmasses migrate. His book, *Origin of the Continents and Oceans*, appeared in 1915. Wegener today is regarded as the father of the concept called continental drift. Wegener postulated that all landmasses were united in one supercontinent approximately 225 million years ago, during the Triassic period, Figure 8.15b. The fact that spreading ridges and subduction zones are areas of earthquake and volcanic activity provides further evidence for plate tectonics, which by 1968 had become the all-encompassing term for these crustal processes.

18. Define upwelling and describe related features on the ocean floor. Define subduction and explain the process.

The worldwide submarine mountain ranges, called the mid-ocean ridges, were the direct result of upwelling flows of magma from hot areas in the upper mantle and asthenosphere. When mantle convection brings magma up to the crust, the crust is fractured and new seafloor is formed, building the ridges and spreading laterally. When continental crust and oceanic crust collide, the heavier ocean floor will dive beneath the lighter continent, thus forming a descending subduction zone (Figure 8.13). The world's oceanic trenches coincide with these subduction zones and are the deepest features on Earth's surface.

Notes on Earth's interior. The depth of convection currents in the mantle is still being investigated. Indirect evidence suggests that there are influences as deep as the Gutenberg discontinuity at the outer core-mantle boundary. Some of the undulating peaks and valleys at that transitional boundary are perhaps triggering some mantle motion. Principal regions of heating and diapir (upward moving hot plumes) formation are still probably within 300 km (185 mi) of the surface. See: Jason Phipps Morgan and Peter M. Shearer, "Seismic constraints on mantle flow and topography of the 660-km discontinuity: evidence for whole-mantle convection," *Nature*, 365, October 1993: 506–11, among many published studies in *Nature* and *Science*.

The lower mantle is thought to be almost uniform in composition and distinct from the upper mantle, containing high-density minerals of iron, magnesium, and silicates, with some calcium and aluminum—known as silicate perovskite. However, much scientific speculation surrounds the composition of the lower mantle. Some scientists think the upper and lower mantle remain separate and distinct. Others think that the 670 km transition zone presents no barrier to mixing and that convection occurs throughout the mantle.

Various metaphors can be used here: the continental plates are barges or rafts floating on the denser asthenosphere. I will sometimes ask if any students have soft-cover textbooks that I can use for a demonstration. Then, standing next to a world physiographic map or "The Ocean Floor" mural map, I use one book as the Nazca plate and the other as the South American plate. As the plates migrate and collide (book spine to spine), I allow the sea-floor book to subduct beneath the continental book. The uplifted edge of the one book becomes the Andes crest.

The original Moho project, designed to drill through oceanic crust, was abandoned in the 1960s. The Russians have been working on the deepest penetration of the crust since 1970—the Kola well under the control of the Russian Interdepartmental Council for the Study of Earth's Interior and Superdeep Drilling. The Russians have started several deep wells. Depths are now approaching 12,000 m (39,400 ft, 7.5 mi). The deepest well in the United States is over 9000 m (29,529 ft). See: Kozlovsky, Ye. A., "The World's Deepest Well," *Scientific American* (December 1984): pp. 98–104.

19. What was Pangaea? What happened to it during the past 225 million years?

See the sequence of illustrations in Figure 8.15, a through e. The supercontinent of Pangaea and its subsequent breakup into today's continents represents only the last 225 million years of Earth's 4.6 billion years, or only the most recent 1/23 of Earth's existence.

20. Characterize the three types of plate boundaries and the actions associated with each type.

The boundaries where plates meet are clearly dynamic places. Divergent boundaries are characteristic of sea-floor spreading centers, where upwelling material from the mantle forms new seafloor, and crustal plates are spread apart. Convergent boundaries are characteristic of collision zones, where areas of continental and/or oceanic crust collide. These are zones of compression. Transform boundaries occur where plates slide laterally past one another at right angles to a sea-floor spreading center, neither diverging nor converging, and usually with no volcanic eruptions.

• *Portray* the pattern of Earth's major plates and *relate* this pattern to the occurrence of earthquakes, volcanic activity, and hot spots.

21. What is the relation between plate boundaries and volcanic and earthquake activity?

Plate boundaries are the primary location of Earth's earthquake and volcanic activity, and the correlation of these phenomena is an important aspect of plate tectonics because they are produced by plate/asthenosphere interactions at these boundaries. Earthquakes and volcanic activity are discussed in more detail in the next chapter, but their general relationship to the tectonic plates is important to point out here. Earthquake zones and volcanic sites are identified on the world plate map in Figure 8.18.

22. What is the nature of motion along a transform fault? Name a famous example of such a fault.

Transform boundaries occur where plates slide laterally past one another at right angles to a sea-floor spreading center (Figure 8.15e). These plates are not diverging or converging, and there is usually no volcanic activity associated with transform boundaries. A famous example of a transform fault is the San Andreas fault; see Figures 9.11c, 9.12, and 9.13.

23. How is the Hawaiian–Emperor chain of islands and seamounts an example of plate motion and hot spot activity? Correlate your answer with the ages of the Hawaiian Islands.

See Figure 8.19. Note the active island-forming eruptions are over the hot spot in the southeastern portion of the big island of Hawai'i. The newest island named Lö'ihi is still rising from beneath the ocean and will take another 10,000 years or more to be hit with tropical sunlight. The Hawaiian Island chain gets progressively older as you move away from the hot spot—as the Pacific Ocean plate moves northwestward. Note that Midway, produced by eruptions from this same hot spot, is 27.7 million years old and that the ages get progressively older as you move to the northwest. This is evidence of a stationary hot spot injecting materials into an overlying plate in motion.

Overhead Transparencies

Tectonics, Earthquakes, and Volcanoes

The chapter-opening illustration, "The Ocean Floor," is a bridge between Chapter 8 and this chapter. The plate map in Figure 8.16 and the gravity anomaly image in Figure 8.17 (both on p. 274) and the map of earthquake and volcanic occurrence in Figure 8.19 (p. 271), as well the world structural map in Figure 9.15 (p. 277), all correlate within this opening illustration. The text along the bottom of the ocean floor map is a tour of the map and students should read the information and locate the places described on the map.

Tectonic activity has repeatedly deformed, recycled, and reshaped Earth's crust during its 4.6-billion-year existence. The principal tectonic and volcanic zones lie along plate boundaries. The arrangement of continents and oceans, the origin of mountain ranges, and the locations of earthquake and volcanic activity are all the result of dynamic endogenic processes. In this chapter we build the continental crust as a basis of world structural regions in Figure 9.18, p. 301. The chapter ends with a discussion of the oftentimes dramatic earthquakes and volcanoes.

Outline Headings and Key Terms

The first-, second-, and third-order headings that divide Chapter 9 serve as an outline for your notes and studies. The key terms and concepts that appear **boldface** in the text are listed here under their appropriate heading in **bold italics**. All these highlighted terms appear in the text glossary. Note the check-off box (❏) so you can mark your progress as you master each concept. Your students have this same outline in their *Student Study Guide*. The ✆ icon indicates that there is an accompanying animation or other resource on the CD.

The outline headings for Chapter 9:

The Ocean Floor
Earth's Surface Relief Features
    ❏ *relief*

    ❏ *topography*
Crustal Orders of Relief
    First Order of Relief
        ❏ *continental landmasses*
        ❏ *ocean basins*
    Second Order of Relief
    Third Order of Relief
    Hypsometry
    Earth's Topographic Regions
Crustal Formation Processes
✆ Terrane Formation
    Continental Shields
        ❏ *continental shield*
    Building Continental Crust
    Terranes
        ❏ *terranes*
Crustal Deformation Processes
    Folding and Broad Warping
✆ **Folds, anticlines, and synclines**
✆ **Fault types**
✆ **Transform Faults, Plate Margins**
        ❏ *folding*
        ❏ *anticline*
        ❏ *syncline*
    Faulting
        ❏ *faulting*
        ❏ *earthquake*
    Normal Fault
        ❏ *normal fault*
    Reverse (Thrust) Fault
        ❏ *reverse fault*
        ❏ *thrust fault*
    Strike-Slip Fault
        ❏ *strike-slip fault*
    Faults in Concert
        ❏ *horst*
        ❏ *graben*
Orogenesis (Mountain Building)
✆ **Plate Boundaries Notebook**
        ❏ *orogenesis*
    Types of Orogenies
        Grand Tetons and the Sierra Nevada

The URLs related to this chapter of *Elemental Geosystems* can be found at
http://www.prenticehall.com/christopherson

Key Learning Concepts

After reading the chapter and using the Student Study Guide, the student should be able to:

• *Describe* first, second, and third orders of relief and *relate* examples of each from Earth's major topographic regions.
• *Describe* the several origins of continental crust and *define* displaced terranes.
• *Explain* compressional processes and folding; *describe* four principal types of faults and their characteristic landforms.
• *Relate* the three types of plate collisions associated with orogenesis and *identify* specific examples of each.
• *Explain* the nature of earthquakes, their measurement, and the nature of faulting.
• *Distinguish* between an effusive and an explosive volcanic eruption and *describe* related landforms using specific examples.

Annotated Chapter Review Questions

• *Describe* **first, second, and third orders of relief and** *relate* **examples of each from Earth's major topographic regions.**

1. **How does the ocean floor map (chapter-opening illustration) exhibit the principles of plate tectonics? Briefly analyze.**

The illustration that opens this chapter is a striking representation of Earth with its blanket of water removed. The scarred ocean floor is clearly visible, its sea-floor spreading centers marked by over 64,000 km (40,000 mi) of oceanic ridges, its subduction zones indicated by deep oceanic trenches, and its transform faults stretching at angles between portions of oceanic ridges. The correlation with various figures in Chapters 8 and 9 is clearly visible.

An ocean floor illustration is available from the National Geographic Society (17th and M Streets N.W., Washington, D.C. 20036; 1-800-638-4077). A world physical map without water or ice titled "World Ocean Floor" (#02683, Mercator projection) costs under $10 and is almost 1.5 m (5 ft) wide; although not the Marie Tharp version, it serves well as a reference wall map. NGS has also prepared a new rendition in their *Atlas of the World*, 7th edition, 2000.

2. **What is meant by an "order of relief"? Give an example from each order.**

Geographers group the landscape's topography into three orders of relief. These orders classify landscapes by scale, from vast ocean basins and continents down to local hills and valleys. The first order of relief consists of continental platforms and oceanic basins. Examples of first order features would be the Pacific Ocean basin and the African continent. Intermediate landforms are considered to be second orders of relief, such as continental masses, mountain masses, plains and lowlands. A few examples are the Alps, Canadian and American Rockies, west Siberian lowland, and the Tibetan Plateau. Great rock cores, or shields, are second order features, see Figure 9.4. In ocean basins, second order features include rises, slopes, abyssal plains, mid-ocean ridges, and submarine trenches. A few examples of these are the Sohm Abyssal Plain, the Mid-Atlantic Ridge, and the Peru-Chile Trench. Third order features are the most detailed forms of relief, consisting of individual mountain, cliffs, valleys, and other landforms of smaller size, such as Mount Diablo in California and Ayres Rock in Australia.

3. Explain the difference between relief and topography.

Relief refers to vertical elevation differences in the landscape. Examples include the low relief of Nebraska and high relief in the Himalayas. Topography is the term used to describe Earth's overall relief, its changing surface form, effectively portrayed on topographic maps.

• *Describe* the several origins of continental crust and *define* displaced terranes.

4. What is a craton? Relate this structure to continental shields and platforms, and describe these regions in North America.

All continents have a nucleus of old crystalline rock on which the continent grows. Cratons are the cores, or heartland regions, of the continental crust. They generally are low in elevation and old (Precambrian, more than 570 million years in age). Those regions where various cratons and ancient mountains are exposed at the surface are called continental shields. Figure 9.4 shows the principal areas of shield exposure.

5. What is a migrating terrane, and how does it add to the formation of continental masses?

Each of Earth's major plates is actually a collage of many crustal pieces acquired from a variety of sources. Accretion, or accumulation, has occurred as crustal fragments of ocean floor, curved chains (or arcs) of volcanic islands, and other pieces of continental crust

have been swept aboard the edges of continental shields. These migrating crustal pieces, which have become attached to the plates, are called terranes. At least 25% of the growth of western North America can be attributed to the accretion of terranes since the early Jurassic period (190 million years ago). A good example is the Wrangell Mountains, which lie just east of Prince William Sound and the city of Valdez, Alaska. The Wrangellia terranes—a former volcanic island arc and associated marine sediments from near the equator—migrated approximately 10,000 km (6200 mi) to form the Wrangell Mountains and three other distinct areas along the western margin of the continent (Figure 9.6).

6. Briefly describe the journey and destination of the Wrangellia Terrane.

The Wrangellia Mountains, which lie just east of Prince William Sound and the city of Valdez, Alaska, are made up of a former volcanic arc and associated marine sediments from near the equator. These terranes migrated approximately 10,000 km (6200 mi) to form distinct formations along the western margin of North America.

• *Explain* compressional processes and folding; *describe* four principal types of faults and their characteristic landforms.

7. Diagram a simple folded landscape in cross section, and identify the features created by the folded strata.

See Figure 9.8. Note the arrow from the illustration to the corresponding photo.

8. Define the four basic types of faults. How are faults related to earthquakes and seismic activity?

See Figure 9.11 and related text section. A type of reverse fault where the fault plane is at a low angle is a thrust fault, the fourth type. When rock strata are strained beyond their ability to remain a solid unit, they fracture, and one side is displaced relative to the other side in a process known as faulting. Thus, fault zones are areas of crustal movement. At the moment the fault line shifts, a sharp release of energy occurs, called an earthquake or quake.

9. How did the Basin and Range Province evolve in the western United States? What other examples exist of this type of landscape?

The Basin and Range Province, in the interior western United States, experienced tensional forces caused by uplifting and thinning of the crust, which cracked the surface to form aligned pairs of normal faults and a distinctive landscape (please refer to Figure

12.23 in Chapter 12). The term horst is applied to upward-faulted blocks; graben refers to downward-faulted blocks. Examples of horst and graben landscapes include the Great Rift Valley of East Africa (associated with crustal spreading), which extends northward to the Red Sea, and the Rhine graben through which the Rhine River flows in Europe. See Figure 9.14 for illustration.

_____

• *Relate* the three types of plate collisions associated with orogenesis and *identify* specific examples of each.

10. Define orogenesis. What is meant by the birth of mountain chains?

Orogenesis literally means the birth of mountains (oros comes from the Greek for mountain). An orogeny is a mountain-building episode that thickens continental crust. It can occur through large-scale deformation and uplift of the crust in episodes of continental plate collision such as the formation of the Himalayan mountains from the collision of India and Asia. It also may include the capture of migrating terranes and cementation of them to the continental margins, and the intrusion of granitic magmas to form plutons. These granite masses often are exposed following uplift and removal of overlying materials. Uplift is the final act of the orogenic cycle. Earth's major chains of folded and faulted mountains, called orogens, bear a remarkable correlation to the plate tectonics model.

11. Name some significant orogenies.

Major orogens include: the Rocky Mountains, produced during the Laramide orogeny (40–80 million years ago); the Sierra Nevada in the Sierra Nevadan orogeny (35 million years ago, with older batholithic intrusions dating back 130–160 million years); the Appalachians and the Valley and Ridge Province formed by the Alleghany orogeny (250–300 million years ago, preceded by at least two earlier orogenies); and the Alps of Europe in the Alpine orogeny (20–120 million years ago and continuing to the present, with many earlier episodes). See the geologic time scale in Figure 8.1, where these are listed.

12. Identify on a map Earth's two large mountain chains. What processes contributed to their development?

The mountain chain that includes the Andes, the Sierra of Central America, the Rockies, and other western mountains was created from the collision between oceanic plates and continental plates. The collision of the Nazca plate with the South American plate, creating the Andes, is a good example of subduction causing folded sedimentary formations, with intrusions of magma forming granitic plutons characteristic of explosive volcanism. Another mountain chain is the Himalayas, created by the collision of the India plate with the Eurasian plate 45 million years ago. This is an example of continental plate-continental plate collision.

13. How are plate boundaries related to episodes of mountain building? Explain how different types of plate boundaries produce differing orogenic episodes and differing landscapes.

Figure 9.16 illustrates the plate-collision pattern associated with each type of orogenesis and points out an actual location on Earth where each mechanism is operational. Shown in (a) is the oceanic plate-continental plate collision type of orogenesis. This occurred along the Pacific coast of the Americas and has formed the Andes, the Sierra of Central America, the Rockies, and other western mountains. Shown in (b) is the oceanic plate-oceanic plate collision, where two portions of oceanic crust collide. This has formed the chains of island arcs and volcanoes that continue from the southwestern Pacific to the western Pacific, the Philippines, the Kurils, and portions of the Aleutians. Shown in (c) is the continental plate-continental plate collision, which occurs when two large continental masses collide. Here the orogenesis is quite mechanical; large masses of continental crust are subjected to intense folding, overthrusting, faulting, and uplifting. As mentioned earlier, the collision of India with the Eurasian landmass produced the Himalayan Mountains.

14. Relate tectonic processes to the formation of the Appalachians and the Alleghany orogeny.

The old, eroded fold-and-thrust belt of the eastern United States contrasts with the younger mountains of the western portions of North America. As noted, the Alleghany orogeny followed at least two earlier orogenic cycles of uplift and the accretion of several captured terranes. (In Europe this is called the Hercynian orogeny.) The original material for the Appalachians and Valley and Ridge Province resulted from the collisions that produced Pangaea.

Again, having the American Geological Institute (AGI) *Glossary of Geology*, 3rd ed., is useful. The mountain-building event, the Alleghany orogeny, which deformed rocks of the Valley and Ridge province and the Allegheny Plateau are spelled with an "a" and an "e" respectively. H.P. Woodward (1957, 1958) introduced the term's spelling (see Figure 9.18 for illustration).

_____

• *Explain* the nature of earthquakes, their measurement, and the nature of faulting.

15. Describe the differences in human response between the two earthquakes that occurred in China in 1975 and 1976.

In the Liaoning Province of northeastern China, ominous indications of possible tectonic activity began in 1970. Foreboding symptoms included land uplift and tilting, increased numbers of minor tremors, and changes in the region's magnetic field—all of this after almost 120 years of quiet. These precursors of coming tectonic events continued for almost five years before Chinese scientists took the bold step of forecasting an earthquake. Finally, on 4 February 1975 at 2:00 P.M., some 3 million people were evacuated in what turned out to be a timely manner; the quake struck at 7:36 P.M., within the predicted time frame. Ninety percent of the buildings in the city of Haicheng were destroyed, but thousands of lives were saved, and success was proclaimed—an earthquake had been forecasted and preparatory action taken for the first time in history.

Only 17 months later, at Tangshan in the northeastern province of Hebei (Hopei), an earthquake occurred on 28 July 1976 without any warning precursors from which a forecast could be prepared. Consequently, this quake destroyed 95% of the buildings and 80% of the industrial structures, severely damaged more than half the bridges and highways, and killed a quarter of a million people! The jolt was strong enough to throw people against the ceilings of their homes. An old, undetected fault had ruptured and offset 1.5 m (5 ft) along an 8 km (5 mi) stretch through Tangshan, devastating large areas just 145 km (90 mi) southeast of Beijing, China's capital city.

16. Compare the North Anatolian, Turkey, and San Andreas, California, fault systems. Discuss their similarities and differences.

See Figure 9.13, p. 290. By orienting California with north to the lower right of the figure, the similarities between the two strike-slip, right-lateral faults in Turkey and California can be seen. Note the motion arrow along each fault system and that neither is a single fault line but rather many faults aligned as a fault system.

17. Differentiate between the Mercalli and moment magnitude and amplitude scales. How are these used to describe an earthquake? Why has the Richter scale been updated and modified?

Earthquake intensity is rated on the arbitrary Mercalli scale, a Roman numeral scale from I to XII representing "barely felt" to "catastrophic total destruction." It was designed in 1902 and modified in 1931 to be more applicable to conditions in North America. Intensity scales are useful in classifying and describing terrain, construction, and local damage conditions following an earthquake.

Earthquake magnitude is estimated according to a system originally designed by Charles Richter in 1935. In this method, the amplitude of a seismic wave is recorded on a seismograph located at least 100 km (62 mi) from the epicenter of the quake. That measure is then charted on the Richter scale, which is open-ended and logarithmic; that is, each whole number on the scale represents a 10-fold increase in the measured wave amplitude. Translated into energy, each whole number demonstrates a 31.5-fold increase in the amount of energy released.

Today, the Richter scale has been improved and made more quantitative. The need for revision was because at higher magnitudes, the scale did not properly measure or differentiate between quakes of high intensity. The moment magnitude scale, in use since 1993, is considered more accurate than Richter's scale for large earthquakes. Moment magnitude considers the amount of fault slippage produced by the earthquake, the size of the surface or subsurface area that ruptured, and the nature of the materials that faulted, including how resistant they were to failure.

18. What is the relationship between an epicenter and the focus of an earthquake? Give examples from the Loma Prieta, California, and Kobe, Japan earthquakes.

The subsurface area along a fault plane, where the motion of seismic waves is initiated, is called the focus, or hypocenter (Figure 9.21 details this for the Loma Prieta earthquake). The area at the surface directly above this subsurface location is the epicenter. Shock waves produced by an earthquake radiate outward from both the focus and epicenter. An aftershock may occur after the main shock, sharing the same general area of the epicenter. A foreshock is also possible preceding the main shock. Kobe is described in News Report 9.2.

19. What local soil and surface conditions in San Francisco severely magnify the energy felt in earthquakes?

If the ground is unstable, distant effects can be magnified, as they were in San Francisco and Mexico. In the Marina District of San Francisco and at the Cypress Freeway structure in Oakland, areas of landfill magnified earthquake energy in 1989, resulting in catastrophic damage. Soils tend to liquefy when shaken as groundwater migrates toward the surface.

Likewise, Mexico City, currently the world's second largest, is positioned on the soft, moist sediments of an ancient lake bed. On 19 September

1985, during rush-hour traffic, two major earthquakes (8.1 and 7.6 on the Richter scale) struck 400 km (250 mi) southwest of Mexico City. The epicenter was on the seafloor off Mexico and Central America, and the old lake bed beneath Mexico City magnified the shock waves by more than 500%.

20. How do the elastic-rebound theory and asperities help explain the nature of faulting?

The elastic-rebound process is described by the elastic-rebound theory. Generally, two sides along a fault appear to be locked by friction, resisting any movement despite the powerful motions of adjoining pieces of crust. This stress continues to build strain along the fault surfaces, storing elastic energy like a wound-up spring. When movement finally does occur as the strain build-up exceeds the frictional lock, energy is released abruptly, returning both sides of the fault to a condition of less strain. Think of the fault plane as a surface with irregularities that act as sticking points, preventing movement, similar to two pieces of wood held together by drops of glue rather than an even coating of glue. Research scientists at the USGS and the University of California have identified these small areas of high strain as asperities. They are the points that break and release the sides of the fault.

21. Describe the San Andreas fault and its relationship to ancient sea-floor spreading movements along transform faults.

The San Andreas is a good example of the evolution of a spreading center overridden by an advancing continental plate (Figure 9.12a). In the figure, the East Pacific Rise developed as a spreading center with associated transform faults (a), while the North American plate was progressing westward after the breakup of Pangaea. Forces then shifted the transform faults toward a northwest-southeast alignment along a weaving axis (b). Finally, the western margin of North America overrode those shifting transform faults (c). In relative terms, the motion along the fault is right lateral, whereas in absolute terms, the North American plate is still moving westward.

I recommend that you obtain a copy of a USGS book which contains great color maps and schematics, photographs, and a series of articles on the San Andreas fault: *The San Andreas Fault System*, California, USGS Robert E. Wallace, ed., Professional Paper 1515, Washington: USGS, 1990, 283 pp.

22. How does the seismic gap concept relate to expected earthquake occurrences? Are any gaps correlated with earthquake events in the recent past? Explain.

One approach to earthquake prediction is to examine the history of each plate boundary and determine the frequency of earthquakes in the past, a study called paleoseismology. Paleoseismologists construct maps that provide an estimate of expected earthquake activity. An area that is quiet and overdue for an earthquake is termed a seismic gap, that is, an area that forms a gap in the earthquake occurrence record and therefore a place that possesses accumulated strain. The area along the Aleutian Trench subduction zone had three such gaps until the great 1964 Alaskan earthquake (8.6 on the Richter scale) filled one of them.

23. What do you see as the biggest barrier to effective earthquake prediction?

In keeping with the perception and political theme of Figure 9.21, p. 301, one of the initial efforts following the 1906 quake by city officials was to put a proper "spin" on the earthquake itself. Public relation efforts centered on blaming fire as the main destructive agent—a preventable danger—rather than the unknown danger of future earthquakes that could destroy the growth and development of San Francisco in the century ahead. An archivist from the San Francisco Library documented these efforts.

We should not use terms such as earthquake prevention or even use of the words "earthquake-proof" when referring to earthquakes. These terms are not used in geoscience nor in civil engineering disciplines; rather they are more frequently used in the political arena. Detailed discussion of hazards and hazard perceptions appear in the writings of Ian Burton (University of Toronto) and Robert W. Kates (Clark University).

Figure 9.21, "Socio-economic impacts and adjustment to earthquakes," can be a powerful teaching tool relative to hazard perception and the political-economic arena. Summarized:

A valid and applicable generalization seems to be that humans are unable or unwilling to perceive hazards in a familiar environment. In other words, we tend to feel secure in our homes and communities, even if they are sitting on a quiet fault zone. Such an axiom of human behavior certainly helps explain why large populations continue to live and work in earthquake-prone settings in developed countries. Similar questions also can be raised about populations in areas vulnerable to other disasters in areas vulnerable to floods, droughts, coastal storm surge, or hurricanes.

Given this axiom, try substituting other natural and human hazards in the "earthquake prediction" box in Figure 9.21. You can place floodplain zoning, global warming prediction, stratospheric ozone depletion, prevention of toxic waste dumping by industry, air bags in cars, cigarette smoking—whatever—in the box and read the litany of socioeconomic impacts as assessed by related special interests. On a global scale, try placing "world peace and disarmament" in the box and the students can see the economic impact posed. This illustrates the oft-cited "that strategy will hurt jobs and the economy" or, "that just can't be valid—it would be too damaging to the economy." I have included this chart in the text because of this versatility and the political and economic reality it portrays. The irony is that proper planning, zoning, and proactive tactics would actually save lives and money in the long run.

Notes on human-caused earthquakes. Las Vegas shook and quaked on command, not from the "action" in the city, but from human-made earthquakes induced by underground testing of nuclear warheads at the Nevada Test Site north of Las Vegas. The first indication that humans could create earthquake-like shock waves came from these tests in the 1950s through the 1980s.

Denver, Colorado, had not experienced earthquakes in 90 years, at least nothing greater than a 3.0 Richter. Early in 1962, quakes mysteriously began to occur and continued for four years, some strong enough to cause damage. The spatial distribution of the quake epicenters were north of Denver. In late 1965, the announcement was made that the U.S. Army created the recent quake activity. The Army was pumping chemical wastes into a 3658 m (12,000 ft) deep well. These wastes were causing fluid pressure in basement rock that increased sliding along fracture zones—fluids under pressure were "lubricating" shear zones. The movements were felt in Denver as earthquakes. Pumping records for the wells confirmed the correlation; the Army reluctantly halted the pumping. A confirming experiment was conducted at an abandoned oil field near Rangely, Colorado, between 1969 and 1973. Water was pumped in and subsequently out of wells in the field, which was specially instrumented for the tests. The USGS reported a strong correlation between fluid injection, producing fluid pressure, and earthquakes. When the water was withdrawn, the earthquakes ceased.

The knowledge of the role of fluid pressure and earthquakes is helping us to understand the safety of large dams and their possible seismic effects. In August 1975, the largest earth fill dam in North America was rocked by a 5.7 Richter earthquake that created little damage but much interest. Evidently, fluid pressure related to the rapid filling of the reservoir the previous spring caused the release of accumulated strain along a susceptible fault zone. The Oroville Dam in east central California, 236 m (775 ft) high with a capacity of 4300 million m3, was filled by early 1969. Although conclusions are still unresolved, Dr. Bruce Bolt of the University of California later wrote:

Undoubtedly, it [the nearby reservoir] sent a pressure pulse through the water in the rocks of the crust. Perhaps, as the pressure pulse spread out by percolation through the crustal rocks nearby, it eventually reached a weak place along an already existing fault zone....it may have been sufficient to open microcracks just enough to allow fault slip (*Earthquakes: A Primer*, W.H. Freeman, 1978, pp. 129–130).

This event added an important parameter to help determine the behavior of a dam and aspects of dam safety. Estimates are that 5 percent of the world's large dams are candidates for seismic effects, although fewer than a dozen cases have been thoroughly studied. On the other hand, other researchers have found evidence that filling a reservoir creates a "loading" situation that actually reduces earthquake frequency in the area of the reservoir body.

———————

• *Distinguish* between an effusive and an explosive volcanic eruption and *describe* related landforms, using specific examples.

24. What is a volcano? In general terms, describe some related features.
A volcano forms at the end of a central vent or pipe that rises from the asthenosphere through the crust into the volcanic mountain, usually forming a crater, or circular surface depression at the summit. Magma rises and collects in a magma chamber deep below the volcano until conditions are right for an eruption. Other features related to volcanic activity are: calderas, large basin-shaped depressions formed when summit material on a volcanic mountain collapses inward after eruption or loss of magma; cinder cones, small cone-shaped hills with a truncated top formed from cinders that accumulate during moderately explosive eruptions; and, shield volcanoes, that are created by effusive volcanism, similar in shape to a shield of armor lying face up on the ground.

25. Where do you expect to find volcanic activity in the world? Why?
See Figure 9.24 and insets 9.25 to 9.30. Also, review the plate tectonic map and volcanic activity in Figure 8.18. The location of volcanic mountains on Earth is a function of plate tectonics and hot spot activity. Volcanic activity occurs in three areas: along subduction boundaries at continental plate-oceanic plate or oceanic plate-oceanic plate convergence; along sea-floor spreading centers on the ocean floor and areas of rifting on continental plates; and at hot spots, where individual plumes of magma rise through the crust.

26. Compare effusive and explosive eruptions. Why are they different? What distinct landforms are produced by each type? Give examples of each.

Effusive eruptions are the relatively gentle eruptions that produce enormous volumes of lava on the seafloor and in places like Hawaii. Direct eruptions from the asthenosphere produce a low-viscosity magma that is very fluid and yields a dark, basaltic rock (less than 50% silica and rich in iron and magnesium). Gases readily escape from this magma because of its texture. A typical mountain landform built from effusive eruptions is gently sloped, gradually rising from the surrounding landscape to a summit crater, similar in outline to a shield of armor lying face up on the ground, and is therefore called a shield volcano.

Volcanic activity along subduction zones produces the well-known explosive volcanoes. Magma produced by the melting of subducted oceanic plate and other materials is thicker (more viscous) than magma from effusive volcanoes; it is 50–75% silica and high in aluminum. Consequently, it tends to block the magma conduit inside the volcano, allowing pressure to build and leading to an explosive eruption. The term composite volcano, or composite cone, is used to describe explosively formed mountains. Composite volcanoes tend to have steep sides and are more conical than shield volcanoes, and therefore they are also known as composite cones.

27. Describe several recent volcanic eruptions.

See the listing at the beginning of the section on page 300. Recent violent eruptions through 1999 remind us of Earth's internal energy: Mount Pinatubo and Mayon volcano (Philippines, 1991 and 1993 respectively), Mount Unzen (Japan, 1991), Mount Hudson (Chile, 1992), Mount Spurr and Mount Redoubt (Alaska, 1992), and Galeras volcano (Colombia, 1993), Rabual Caldera (Papua, New Guinea, 1994), Mount Etna (Italy, 1995), and Nyiragongo (Congo, 2002). Consult Table 9.3, as well as the Volcanism section in Chapter 9, for lists of other examples.

Overhead Transparencies

182.	CO 9	*Floor of the Oceans*, 1975, (Marie Tharp)
183.	9.2	Earth's hypsometry
184.	9.3	Earth's topographic regions map and legend
185.	9.4 b	Continental shields map
186.	9.5	Crustal formation processes
187.	9.7 a, b, c	Three kinds of stress, strain, and resulting surface expressions
188.	9.8 a, b, c	Folded landscapes
189.	9.10 a, b, c, d	Upwarped dome, structural basin
190.	9.11 a, b, c	Three types of faults
191.	9.12 a, b, c	San Andreas fault formation; southern Calif. inset map and fault line photo
192.	9.13	Comparison of strike-slip faults in Turkey and California
193.	9.14a, b	Horst and graben; Red Sea from orbit
194.	9.15	European Alps satellite image and locator map
195.	9.16 a, b, c	Three types of plate convergence
196.	9.17	Tilted-fault block Tetons, photo and schematic
197.	9.18a, b	Satellite image of the Appalachian region
198.	9.19	World structural regions and major mountain systems; *Landsat* image of North and South America
199.	9.20a, b, c	Illustration of fault-plane solution of the Loma Prieta earthquake
200.	9.21	Socio-economic impacts to an earthquake adjustment
201.	9.22	Tectonic settings of volcanic activity
202.	9.30 and 9.31	Shield and composite volcanoes compared (top); a composite volcano (bottom)
203.	N.R. 9.3.1	CO_2 killing the forest; locator map for Long Valley, CA
204.	F.S. 9.1.2	Mount St. Helens

Weathering, Karst Landscapes, and Mass Movement

Chapter 10 begins the treatment of the exogenic system. Chapters 8 and 9 covered the endogenic system. As the landscape is formed, a variety of exogenic processes simultaneously operate to wear it down. The endogenic system builds and creates initial landscapes, while the exogenic system works towards low relief, little change, and the stability of sequential landscapes.

Students can be asked if they have noticed highways in mountainous and cold climates that appear rough and broken. Roads that experience freezing weather seem to pop up in chunks each winter. Or maybe they have seen older marble structures, such as tombstones, etched and dissolved by rainwater. Similar physical and chemical weathering processes are important to the overall reduction of the landscape and the release of essential minerals from bedrock. A simple examination of soil gives evidence of weathered mineral grains from many diverse sources. In addition, mass movement of surface material rearranges landforms, providing often-dramatic reminders of the power of nature.

Outline Headings and Key Terms

The first-, second-, and third-order headings that divide Chapter 10 serve as an outline for your notes and studies. The key terms and concepts that appear **boldface** in the text are listed here under their appropriate heading in ***bold italics***. All these highlighted terms appear in the text glossary. Note the check-off box (❏) so you can mark your progress as you master each concept. Your students have this same outline in their *Student Study Guide*. The ✎ icon indicates that there is an accompanying animation or other resource on the CD.

The outline headings for Chapter 10:
Landmass Denudation
 ❏ *geomorphology*
 ❏ *denudation*
 ❏ *differential weathering*
Dynamic Equilibrium Approach to Landforms
 ❏ *dynamic equilibrium model*
 ❏ *geomorphic threshold*
Slopes
 ❏ *slopes*
Weathering Processes
 ❏ *weathering*
 ❏ *regolith*
 ❏ *bedrock*
 ❏ *sediment*
 ❏ *parent material*
Factors Influencing Weathering Processes
 ❏ *joints*
Physical Weathering Processes
✎ **Physical Weathering**
 ❏ *physical weathering*
Frost Action
 ❏ *frost action*
Crystallization
Pressure-Release Jointing
 ❏ *sheeting*
 ❏ *exfoliation dome*
Chemical Weathering Processes
 ❏ *chemical weathering*
 ❏ *spheroidal weathering*
Hydration and Hydrolysis
 ❏ *hydration*
 ❏ *hydrolysis*
Oxidation
 ❏ *oxidation*
Carbonation and Solution
 ❏ *carbonation*

The URLs related to this chapter of *Elemental Geosystems* can be found at **http://www.prenticehall.com/christopherson**

Key Learning Concepts

After reading the chapter and using the Student Study Guide, the student should be able to:

• *Define* the science of geomorphology.
• *Illustrate* the forces at work on materials residing on a slope.
• *Define* weathering and *explain* the importance of the parent rock and joints and fractures in rock.
• *Describe* frost action, crystallization (salt crystal growth), pressure-release jointing, and the role of freezing water as physical weathering processes.
• *Describe* the susceptibility of different minerals to the chemical weathering processes including hydration, hydrolysis, oxidation, carbonation, and solution.
• *Review* the processes and features associated with karst topography.

• *Portray* the various types of mass movements and *identify* examples of each in relation to moisture content and speed of movement.

Annotated Chapter Review Questions

• *Define* the science of geomorphology.

1. Define geomorphology and describe its relationship with physical geography.

Geomorphology is a science that analyzes and describes the origin, evolution, form, and spatial distribution of landforms. It is an important aspect of the study of physical geography and the understanding of the spatial-physical aspects of landforms.

An interesting example familiar to most is the Grand Canyon of Arizona, an example of form and process: the erosional processes that reduce a landscape are balanced against the resistance of the materials that make up the landscape. Weathering and erosional forces naturally oscillate, especially in the desert, with high rainfall variability coming in episodic thunderstorms. Each rainfall event at the Grand Canyon operates on available slopes and cliffs. The river receives materials and discharges its flow of water and sediment load. The variation in rock resistance is responsible for the cliffs and slopes in the rock: more resistant rock cliffs, less resistant rock slopes.

The tremendous uplift of the Colorado Plateau over the past 10 million years rejuvenated the profile of the Colorado River and facilitated the downwasting of the surrounding landscape. The river drops 670 m (2200 ft) in elevation as it makes its way through the Grand Canyon region. Today, the river is almost 1.6 km (1 mi) below the rim of the canyon; the cliffs and slopes have retreated to the extent that the canyon is over 16 km (10 mi) wide. What is revealed is an incredible history of the entire region, a time span of over 2 billion years of Earth's history. The sea has entered at least seven times, each episode clearly marked by layers of limestone and beach sands hardened (lithified) to sandstone.

Endogenic periods of uplift did not take place in a smooth sequence; quiet periods occurred and are recorded in rock formations by plateau-like terraces called esplanades in the canyon—the Tonto Plateau. Metamorphic rocks in the inner gorge are the roots of old mountains. Imagine, as you look at a photograph of the canyon, the constant and ongoing adjustments that are occurring in each portion of the landscape. Think of the geomorphic thresholds that are breached following thunderstorms and flash floods throughout the canyon. Yet what we witness is a beautiful, quiet, and seemingly motionless landscape that is constantly changing over a grand time scale.

2. Define landmass denudation. What processes are included in the concept?

Denudation is a general term referring to all processes that cause reduction or rearrangement of landforms. The principal denudation processes affecting surface materials include weathering, mass movement, erosion, transportation, and deposition.

3. What is the interplay between the resistance of rock structures and weathering variabilities?

Weathering is greatly influenced by the character of the bedrock: hard or soft, soluble or insoluble, broken or unbroken. The differing resistance of rock, coupled with these variations in the intensity of weathering, result in differential weathering.

Interactions between the structural elements of the land and denudation processes are complex, and represent a constant struggle between internal and external processes. An important question to ask is whether or not this dynamic interplay is progressive, evolving and building landforms in an orderly manner through stages. Do landscapes initially form and subsequently age in graceful stages until they are flat? Or does the interplay of forces fluctuate back and forth across a never achieved steady-state equilibrium? The debates in geomorphology are fueled by the fact that landscapes evolve on a much longer time span than does a human life, or a research grant, or thesis project, or the edition of a textbook. Modern geomorphology has moved away from simple descriptive classifications.

4. Explain how Figure 10.1 is an example of differential weathering.

Delicate Arch is a dramatic example of differential weathering in Arches National Park, Utah. Resistant rock strata at the top of the structure have helped preserve the arch beneath them as surrounding rock was eroded away. The arch stands 15 stories tall!

5. What are the principal considerations in the dynamic equilibrium model?

The balancing act between tectonic uplift and reduction by weathering and erosion, between the resistance of rocks and the ceaseless attack of weathering and erosion, is summarized in the dynamic equilibrium model. A dynamic equilibrium demonstrates a trend over time. According to current thinking, landscapes in a dynamic equilibrium show ongoing adaptations to the ever-changing conditions of rock structure, climate, local relief, and elevation. Endogenic events (such as earthquakes and volcanic eruptions), or exogenic events (such as heavy rainfall or

forest fire), may provide new sets of relationships for the landscape.

• *Illustrate* **the forces at work on materials residing on a slope.**

6. Describe conditions on a hillslope that is right at the geomorphic threshold. What factors might push the slope beyond this point?

As changing conditions provide new sets of relationships for the landscape, the system eventually arrives at a geomorphic threshold, or that point at which the system breaks through to a new set of equilibrium relationships and rapidly realigns landscape materials accordingly. Slopes, as parts of landscapes, are open systems and seek an angle of equilibrium among the forces described here. Conflicting forces work together on slopes to establish an optimum compromise incline that balances these forces. A geomorphic threshold (change point) is reached when any of the conditions in the balance is altered. Many factors could alter the hillside's equilibrium, such as an earthquake, saturation of the regolith, or the building of a house or dam (adding mass). All the forces on the slope then compensate by adjusting to a new dynamic equilibrium.

Slopes seek an angle of equilibrium among the forces described in the text (Figure 10.3). Strahler in 1950 calculated slope equilibrium angles and found great uniformity for the specific area he studied. In other words, slopes exhibit maximum inclines that balance conflicting forces. A geomorphic threshold (change point) is reached if one of the conditions in the balance alters. When this occurs all the forces on the slope compensate by adjusting to a new dynamic equilibrium. A slope is an open system responding to variable inputs and producing variable outputs.

In a generic sense and under most climatic regimes, slopes have the same four identifiable traits: a waxing slope, a free-face scarp, a debris slope, and a waning slope or pediment. Variation in the resistance of rock strata or climatic conditions may alter the conformations in this ideal model. Weathering and mass movement principally work on the upper portions of the slope, while running water predominates on the lower slopes. Figure 13.3b illustrates these principal elements of slope form. The waxing slope is a forming slope of convex shape that is consumed by the enlarging free face scarp, and retreats as the landscape is reduced. The debris slope receives rock fragments and materials from above, with continuously moving water in humid climates carrying away material as it arrives. In arid climates, graded debris slopes persist where moving water is infrequent. And finally, the waning slope, or pediment, has a graded form in a river valley, or is a wide, flat intermittently dry lake bed in arid areas.

7. Given all the interacting variables, do you think a landscape ever reaches a stable, old-age condition? Explain.

In reality, a landscape behaves as an open system, with highly variable inputs of energy and materials. In response to this input of energy and materials, landforms constantly adjust toward a condition of equilibrium. As physical factors fluctuate, the surface constantly responds in search of an equilibrium condition, with every change producing compensating actions and reactions. The balancing act between tectonic uplift and the reduction rates of weathering and erosion, and between the resistance of crust materials and the ceaseless attack of denudation processes, is summarized in the dynamic equilibrium model. A dynamic equilibrium is different from a steady-state equilibrium, which fluctuates around an average. Instead, according to current thinking, landscapes adapt to the ever-changing conditions of rock structure, climate, local relief, and elevation. Various endogenic episodes, such as faulting, or exogenic episodes, such as a heavy rainfall, may episodically provide new sets of relationships for the landscape. According to Davis' evolutionary cyclic model, the landscape eventually evolves into an old erosional surface that he called a peneplain. In reality, a peneplain has never been identified in nature; landscapes do not hold still for such prolonged periods of time.

8. What are the general components of an ideal slope?

Figure 10.3b illustrates basic slope components that vary with conditions of rock structure and climate. Slopes generally feature an upper waxing slope near the top. The convex surface curves downward and grades into the free face below. The presence of a free face indicates an outcrop of resistant rock that forms a steep scarp or cliff.

Downslope from the free face is a debris slope, which receives rock fragments and materials from above. The debris slope grades into a waning slope, concave surface along the base of the slope. This surface of erosional materials gently slopes at a continuously decreasing angle to the valley floor.

A slope is an open system seeking an angle of equilibrium. Conflicting forces work simultaneously on slopes to establish an optimum compromise incline that balances these forces.

9. Relative to slopes, what is meant by an "angle of equilibrium"? Can you apply this to the photograph in Figure 10.2?

Slopes, as parts of landscapes, are open systems and seek an angle of equilibrium among the forces described in the answer to Question 6 above. The recently disturbed hillslope in Figure 10.4 is in the midst of compensating adjustment. The disequilibrium was created by the failure of saturated slopes caused by a heavy thunderstorm downpour.

• *Define* weathering and *explain* the importance of the parent rock and joints and fractures in rock.

10. Describe weathering processes operating on an open expanse of bedrock. How does regolith develop? How is sediment derived?

Rocks at or near Earth's surface are exposed to both physical and chemical weathering processes. Weathering encompasses a group of processes by which surface and subsurface rock disintegrates into mineral particles or dissolves into minerals in solution. Weathering does not transport the weathered materials; it simply generates these raw materials for transport by the agents of wind, water, and gravity. In most areas, the upper surface of bedrock is partially weathered to broken-up rock called regolith. In some areas, regolith may be missing or undeveloped, thus exposing an outcrop of unweathered bedrock. Loose surface material comes from further weathering of regolith and from transported and deposited regolith. This unconsolidated sediment and weathered rock forms the parent material from which soil evolves.

When rock is broken and disintegrated without any chemical alteration, the process is called physical weathering or mechanical weathering. By breaking up rock, physical weathering greatly increases the surface area on which chemical weathering may operate. Chemical weathering refers to actual decomposition and decay of the constituent minerals in rock due to chemical alteration of those minerals. A familiar example of chemical weathering is the eating away of cathedral facades and etchings on tombstones caused by increasingly acid precipitation. Water is essential to chemical weathering and, as shown in Figure 10.10, the chemical breakdown becomes more intense as both temperature and precipitation increase.

11. Describe the relationship between climatic conditions and rates of weathering activities. Is the scale at which we consider weathering processes important?

Important in determining weathering rates are climatic elements, including the amount of precipitation, overall temperature patterns, and any freeze-thaw cycles. Annual precipitation and temperature and the physical (mechanical) or chemical weathering processes in an area are related to patterns of temperature and precipitation. You can see that physical weathering dominates in drier, cooler climates, whereas chemical weathering dominates in wetter, warmer climates. Extreme dryness reduces most weathering to a minimum, as is experienced in desert climates. (Refer to Chapter 7 for the key to these climate designations.) In the hot, wet-tropical and equatorial rain-forest climates, most rocks weather rapidly, and the weathering tends to be deep below the surface. Note the continued weathering at microscale sites noted on the figure.

12. What is the relation among parent rock, parent material, regolith and soil?

Bedrock is the parent rock from which weathered regolith and soils develop. While a soil is relatively youthful, its parent rock is traceable through similarities in composition. This unconsolidated fragmental material, known as sediment, combines with weathered rock to form the parent material from which soil evolves. See Figure 10.4c.

13. What role do joints play in the weathering process? Give an example from one of the illustrations in this chapter.

Joints are fractures or separations in rock that occur without displacement of the sides (as would be the case in faulting). The presence of these usually plane (flat) surfaces greatly increase the surface area of rock exposed to both physical and chemical weathering. An example is found in Figure 10.6, which portrays joint-block separation, a form of physical weathering.

_____

• *Describe* frost action, crystallization (salt crystal growth), pressure-release jointing, and the role of freezing water as physical weathering processes.

14. What is physical weathering? Give an example.

Physical, or mechanical, weathering is the term used when rock is broken and disintegrated without any chemical alteration. By breaking up rock, physical weathering greatly increases the surface area on which chemical weathering can take place. An example of physical weathering is frost action. When water freezes, its volume expands. This creates a powerful mechanical force, which can exceed the tensional strength of rock. Repeated freezing and thawing of water break rocks apart. The work of ice begins in small openings, such as existing joints and fractures, gradually expanding until rocks are split apart.

15. Why is freezing water such an effective physical weathering agent?

Water expands by as much as 9% of its volume as it freezes (see Chapter 5). This expansion creates a powerful mechanical force that can exceed the tensional strength of rock. In a weathering action called frost-wedging, ice crystals grow in preexisting cracks in rock and push the sides apart along joints or fractures. Figures 10.5b and 10.6 show this action on blocks of rock, which causes joint-block separation, although cracking and breaking can be in any shape. The work of ice probably begins in small openings, gradually expanding until rocks are cleaved. Softer supporting rock underneath the slabs already has weathered physically—an example of differential weathering.

16. What weathering processes produce a granite dome? Describe the sequence of events.

As the tremendous weight of overburden is removed from a granitic pluton, the pressure of deep burial is relieved. The granite responds with a slow but enormous heave, and in a process known as pressure-release jointing, layer after layer of rock peels off in curved slabs, thinner toward the top of the formation and thicker toward the sides. As these slabs weather they slip off in a process called spalling, and form an exfoliation dome. This exfoliation process creates arch-shaped and dome-shaped features on the exposed landscape (Figure 10.9). Such domes probably are the largest single weathering features on Earth (in areal extent).

_____

• *Describe* the susceptibility of different minerals to the chemical weathering processes including hydration, hydrolysis, oxidation, carbonation, and solution..

17. What is chemical weathering? Contrast this set of processes to physical weathering.

Chemical weathering is the actual decomposition of minerals in rock. Chemical weathering involves reactions between air, water, and minerals in rock. Minerals may combine with water in chemical reactions (such as carbonation), or carbon dioxide and oxygen from the atmosphere (such as oxidation). Although physical weathering may create greater surface area for further weathering to take place, chemical weathering can dissolve minerals throughout the rock.

An example that demonstrates the difference between physical and chemical weathering is the absorption of water in rocks. In cold climates dominated by physical weathering, the process of frost action takes place when water present in the rock expands with freezing and cracks the rock into smaller pieces. In humid climates, where chemical weathering occurs, water percolates into the rock, a process called hydrolysis, and breaks down the silicate minerals in rock. Hydrolysis dissolves silicate materials, leaving behind resistant minerals, such as quartz.

18. What is meant by the term spheroidal weathering? Describe the spheroidal weathering process?

Spheroidal weathering is an example of the way chemical weathering attacks rock. The sharp edges and corners of rock are rounded as the alteration of minerals progresses through the rock. Joints in the rock offer more surfaces of opportunity for weathering. Water penetrates joints and fractures and dissolves the rock's weaker minerals or cementing materials. The resulting rounded edges are the basis for the name spheroidal weathering.

19. What is hydrolysis? How does it affect rocks? Differentiate this process from hydration? How does the role of water differ?

When minerals chemically combine with water, the process is called hydrolysis. Hydrolysis is a decomposition process that breaks down silicate minerals in rocks. Water is not simply absorbed in hydrolysis but actively participates in chemical reactions to produce different compounds and minerals.

Hydration, meaning "combination with water," involves little chemical change. Water becomes part of the chemical composition of the mineral (such a hydrate is gypsum, which is hydrous calcium sulfate:). When some minerals hydrate, they expand, creating a strong mechanical effect that stresses the rock, forcing grains apart. A cycle of hydration and dehydration can lead to granular disintegration and further susceptibility of the rock to chemical weathering.

20. Iron minerals in rock are susceptible to which form of chemical weathering? What characteristic color is associated with this type of weathering?

Iron minerals in rock are susceptible to oxidation. Oxidation is an example of chemical weathering that occurs when oxygen combines with certain metallic minerals to form oxides. The rusting of iron in rocks or soils produces a reddish-brown stain of iron oxide.

21. With what kind of minerals do carbon compounds react, and under what circumstances does this reaction occur? What is this weathering process called?

Carbon compounds react with carbonic acid, created when water vapor dissolves carbon dioxide. Carbonic acid is strong enough to react with many minerals, especially limestone, in a process known as carbonation. When rainwater attacks formations of limestone, the constituent minerals are dissolved and wash away with the mildly acidic rainwater.

_____

• *Review* the processes and features associated with karst topography.

22. Describe the development of limestone topography. What is the name applied to such landscapes? From what was this name derived?

Limestone is so abundant on Earth that many landscapes are composed of it. These areas are quite susceptible to chemical weathering. Such weathering creates a specific landscape of pitted, bumpy surface topography, poor surface drainage, and well-developed solution channels underground. Remarkable labyrinths of underworld caverns also may develop. These are the hallmarks of karst topography, originally named for the Krs Plateau in Yugoslavia, where these processes were first studied. Approximately 15% of Earth's land area has some developed karst, with outstanding examples found in southern China, Japan, Puerto Rico, Cuba, the Yucatán of Mexico, Kentucky, Indiana, New Mexico, and Florida.

23. Differentiate among sinkholes, karst valleys, and cockpit karst. Within which form is the radio telescope at Arecibo, Puerto Rico?

See Figures 10.13, 10.14, and 10.15. The Arecibo radio telescope (10.15c) operated by Cornell University takes advantage of the symmetrical form of a doline in a complex cockpit topography of numerous dolines.

24. In general, how would you characterize the region southwest of Orleans, Indiana?

The region southwest of Orleans, Indiana, has over 1000 sinkholes in just 2.6 km^2 (1 mi^2). In this area the Lost River, a "disappearing stream," flows more than 12.9 km (8 mi) underground before it resurfaces; its flow diverted from the surface through sinkhole and solution channels. See Figure 10.1b.

25. What are some of the unique erosional and depositional features you find in a limestone cavern?

See Figure 10.17a, and b through f, for examples.

_____

• *Portray* the various types of mass movements and *identify* examples of each in relation to moisture content and speed of movement

26. Define the role of slopes in mass movements, using the terms *angle of repose, driving force, resisting force, and geomorphic threshold*.

All mass movements occur on slopes. The steepness of a slope determines where loose material comes to rest, depending on the size and texture of the grains; this is called the angle of repose. This angle represents a balance of driving and resisting forces.

The driving force in mass movements is gravity, working in conjunction with the weight, size, and shape of the grains or surface material, the degree to which the slope is oversteepened, and the amount and form of moisture available—whether frozen or fluid. The greater the slope angle, the more susceptible the surface material is to mass movement. The resisting force is the shearing strength of slope material, that is, its cohesiveness and internal friction working against mass movement. To reduce shearing strength is to increase shearing stress, which eventually reaches the point at which gravity overcomes friction.

27. What events occurred in the Madison River Canyon in 1959?

Figure 10.18. In the Madison River Canyon near West Yellowstone, Montana, on the Wyoming border, a blockade of dolomite (a magnesium-rich rock in the carbonate group) had held back a deeply weathered and oversteepened slope (with a 40°–60° slope angle) for untold centuries. Then, shortly after midnight on 17 August 1959, an earthquake measuring 7.5 on the Richter scale broke the dolomite structure along the foot of the slope and released 32 million m^3 (1.13 billion ft^3) of mass. This moved downslope at 95 kmph (60 mph), creating gale force winds through the canyon. The material continued more than 120 m (390 ft) up the opposite canyon slope, entombing campers beneath about 80 m (260 ft) of rock and debris.

The mass of material also effectively dammed the Madison River and created a new lake, dubbed Quake Lake. The landslide debris dam established a new temporary base level for the canyon.

28. What are the classes of mass movement? Describe each briefly and differentiate among these classes.

Four basic classifications of mass movement are used: fall, slide, flow, and creep. Each involves the pull of gravity working on a mass until the critical shearing strength is reduced to the point that the mass falls, slides, flows, or creeps downward. The Madison River Canyon event was a type of slide, whereas the Nevado del Ruiz lahar mentioned earlier was a flow. A rockfall is simply a quantity of rock that falls through the air and hits a surface. During a rockfall, individual pieces fall independently, and characteristically form a pile of irregular broken rocks called a talus cone at the base of a steep cliff.

29. Name and describe the type of mudflow associated with a volcanic eruption.

The hot eruption of Nevado del Ruiz in central Colombia, South America, melted about 10% of the ice on the mountain's snowy peak, liquefying mud and volcanic ash and sending a hot mudflow downslope. Such a flow is called a lahar, an Indonesian word referring to flows of volcanic origin. The Mount Saint Helens eruption also produced a lahar in the Toutle River valley.

30. Describe the difference between a landslide and what happened on the slopes of Nevado Huascarán.

A debris avalanche is a mass of falling and tumbling rock, debris, and soil. It is differentiated from a debris slide or landslide by the tremendous velocity achieved by the onrushing materials. These speeds often result from ice and water that fluidize the debris. The extreme danger of a debris avalanche results from these tremendous speeds and lack of warning.

31. What is scarification, and why is it considered a type of mass movement? Give several examples of scarification. Why are humans a significant geomorphic agent?

Large open-pit strip mines—such as the Bingham Copper Mine west of Salt Lake City, the Berkeley Pit in Butte, Montana, and the extensive strip mining for coal in the East and West—are examples of human-induced mass movements, generally called scarification. At the Bingham Copper Mine, a mountain literally was removed. The disposal of tailings and waste material is a significant problem with such large excavations because the tailing piles prove unstable and susceptible to further weathering, mass wasting, or wind dispersal.

R. L. Hooke, Department of Geology and Geophysics, University of Minnesota, used estimates of U.S. excavations for new housing, mineral production (including the three largest—stone, sand and gravel, and coal), and highway construction. Hooke estimated that humans, as a geomorphic agent, annually move 40–45 billion tons (40–45 Gt/yr) of the planet's surface. Therefore, humans exceed natural river sediment transfer (14 Gt/yr), movement through stream meandering (39 Gt/yr), haulage by glaciers (4.3 Gt/yr), movement due to wave action and erosion (1.25 Gt/yr), wind transport (1 Gt/yr), sediment movement by continental and oceanic mountain building (34 Gt/yr), or deep-ocean sedimentation (7 Gt/yr). As Hooke stated,

> Homo sapiens has become an impressive geomorphic agent. Coupling our earth-moving prowess with our inadvertent adding of sediment load to rivers and the visual impact of our activities on the landscape, one is compelled to acknowledge that, for better or for worse, this biogeomorphic agent may be the premier geomorphic agent of our time.

Overhead Transparencies

205.	10.3 a and b	Slope mechanics and form
206.	10.4 a, b, c	Typical hillside cross section, exposed hillside, sand dunes and parent material
207.	10.9a, b, c	Exfoliation examples: White Mtns., close-up of Half Dome, overview of Half Dome
208.	10.11 a – e	Features of karst topography
209.	10.13 a – f	Cavern features
210.	10.14	Madison River landslide: geologic cross-section and photo
211.	10.15a, b, c	Mass movement classes
212.	10.19	Soil creep and effects
213.	F.S. 10.1.1a, b	Vaiont Reservoir disaster

11

River Systems
and Landforms

Earth's rivers and waterways form vast arterial networks that both shape and drain the continents, transporting the by-products of weathering, mass movement, and erosion. To call them Earth's lifeblood is not an exaggerated metaphor, inasmuch as rivers redistribute mineral nutrients important for soil formation and plant growth. Not only do rivers provide us with essential water supplies, but they also receive, dilute, and transport wastes and provide critical cooling water for industry. Rivers have been of fundamental importance throughout human history. This chapter discusses the dynamics of river systems and related landforms that streams produce.

This chapter begins with a discussion of the greatest rivers in the world in terms of flow discharge. Next base level and the essential principles that govern stream flow are considered. The drainage basin is a basic hydrologic unit. A map presents these and the important continental divides for the United States and Canada. With this established, streamflow characteristics, gradient, and deposition are presented as water cascades through the hydrologic system.

The human component is irrevocably linked to streamflow as so many settlements are along river banks and on floodplains. Not only do rivers provide us with essential water supplies, but they also receive, dilute, and transport wastes and provide critical cooling water for industry. Rivers have been of fundamental importance throughout human history.

News Reports in this chapter highlight fascinating aspects of rivers and river management/mismanagement—dam releases to clear sediment out of the Grand Canyon, a disappearing Nile delta, and what once was Bayou Lafourche in Lousiana.

This chapter discusses the dynamics of river systems and their landforms. Our successful interaction with river systems will be central to the sustainability of our economic behavior throughout the new century. Thus we begin our important study of rivers.

A daily check of the news brings home the importance of this chapter. The catastrophic floods along the Mississippi River and its tributaries in 1993 illustrated the risk of floodplain settlement. Total damage from these floods exceeded $30 billion. Nationally, in 2001, some two-thirds of disaster losses were attributable to floods. Tropical storm Allison left $6 billion in damage in its wandering visit to Texas in June 2001, producing most flood damage along occupied floodplains. The 2002 floods in Europe produced estimated losses exceeding U.S. $50 billion. Of course, the 2005 levee breaks caused by poor planning and Hurricane Katrina resulted in catastrophic loss of life and property in New Orleans and will forever be in our memories, as floodwaters engulfed 80% of the city (see the News Report in Chapter 17).

The need to better understand river systems has never been more important—especially for the beleaguered insurance industry. In this spirit we embark on Chapter 11.

Outline Headings and Key Terms

The first-, second-, and third-order headings that divide Chapter 11 serve as an outline for your notes and studies. The key terms and concepts that appear **boldface** in the text are listed here under their appropriate heading in ***bold italics***. All these highlighted terms appear in the text glossary. Note the check-off box (❑) so you can mark your progress as you master each concept. Your students have this same outline in their *Student Study Guide*. The ✪ icon indicates that there is an accompanying animation or other resource on the CD.

The outline headings for Chapter 11:
- ❑ *hydrology*

Fluvial Processes and Landscapes
- ❑ *fluvial*
- ❑ *erosion*
- ❑ *transport*
- ❑ *deposition*
- ❑ *alluvium*

Base Level of Streams
- ❑ *base level*

The Drainage Basin System
- ❑ *drainage basin*
- ❑ *watershed*
- ❑ *sheetflow*

Drainage Divides and Basins
- ❑ *continental divides*

Drainage Basins as Open Systems

An Example: the Delaware River Basin

Drainage Patterns
- ❑ *drainage pattern*

Streamflow Characteristics
- ✪ **Stream processes, Floodplains**
- ✪ **Oxbow lake formation**
- ✪ **Niagara Falls Notebook**
  - ❑ *discharge*

Stream Erosion
- ❑ *hydraulic action*
- ❑ *abrasion*

Stream Transport
- ❑ *dissolved load*
- ❑ *suspended load*
- ❑ *bed load*
- ❑ *traction*
- ❑ *saltation*
- ❑ *aggradation*
- ❑ *braided stream*

Flow and Channel Characteristics

- ❑ *meandering stream*
- ❑ *undercut bank*
- ❑ *point bar*
- ❑ *oxbow lake*

Stream Gradient
- ❑ *gradient*
- ❑ *graded stream*

Nickpoints
- ❑ *nickpoint*

Stream Deposition

✪ **Stream Terrace Formation**

Floodplains
- ❑ *floodplain*
- ❑ *natural levees*

Stream Terraces
- ❑ *alluvial terraces*

River Deltas
- ❑ *delta*
- ❑ *estuary*

Mississippi River Delta

Rivers Without Deltas

Floods and River Management

Floods and Floodplains
- ❑ *flood*

Streamflow Measurement
- ❑ *hydrographs*

Summary and Review

News Reports and Focus Study

News Report 11.1: Scouring the Grand Canyon for New Beaches and Habitats

News Report 11.2: The Nile Delta is Disappearing

News Report 11.3: What Once Was Bayou Lafourche—Analysis of a Photo

Focus Study 11.1: Floodplain Management Strategies

The URLs related to this chapter of *Elemental Geosystems* can be found at **http://www.prenticehall.com/christopherson**

Key Learning Concepts

After reading the chapter and using the Student Study Guide, the student should be able to:

• *Define* fluvial and *outline* the fluvial processes: erosion, transportation, and deposition.

• *Construct* a basic drainage basin model and *identify* different types of drainage patterns, with examples.

• *Describe* the relation among velocity, depth, width, and discharge, and *explain* the various ways that a stream erodes and transports its load.

• *Develop* a model of a meandering stream, including point bar, undercut bank, and cutoff, and *explain* the role of stream gradient in these flow characteristics.

• *Define* a floodplain and *analyze* the behavior of a stream channel during a flood.

• *Differentiate* the several types of river deltas and *detail* each.

• *Explain* flood probability estimates and *review* strategies for mitigating flood hazards.

Annotated Chapter Review Questions

• *Define* the term fluvial and *outline* the fluvial processes: erosion, transportation, and deposition.

1. What role is played by rivers in the hydrologic cycle?

Earth's rivers and waterways form vast arterial networks that both shape and drain the continents, transporting the by-products of weathering, mass movement, and erosion. To call them Earth's lifeblood is not an exaggerated metaphor, inasmuch as rivers redistribute mineral nutrients important for soil formation and plant growth. Not only do rivers provide us with essential water supplies, but they also receive, dilute, and transport wastes and provide critical cooling water for industry. Rivers have been of fundamental importance throughout human history.

2. What are the five largest rivers on Earth in terms of discharge? Relate these to the weather patterns in each area and to regional potential evapotranspiration (moisture demand) and precipitation (moisture supply).

See Table 11.1 for a listing of these rivers. Relative to water budgets, areas of seasonal and annual surplus produce higher discharges.

3. Define the term fluvial. What is a fluvial process?

Stream-related processes are termed fluvial (from the Latin *fluvius*, meaning river). Insolation is the driving force of fluvial systems, operating through the hydrologic cycle and working under the influence of gravity. Denudation by water dislodges, dissolves, or removes surface material as erosional fluvial processes. Thus, streams supply weathered and wasted sediments for transport to new locations, where they are laid down in a process known as deposition.

4. What is the difference between a local and an ultimate base level

John Wesley Powell (1834–1902) was an early director of the U.S. Geological Survey, an explorer of the Colorado River, and a pioneer in understanding the landscape. In 1875 he put forward the idea of **base level**, or a level below which a stream cannot erode its valley further. The hypothetical *ultimate base level* is sea level. You can imagine base level as a surface extending inland from sea level, inclined gently upward, under the continents (Figure 11.3). Ideally, this is the lowest practical level for all denudation processes.

Powell also recognized that not every landscape degraded down to sea level, for other base levels seemed to be in operation. A *local base level*, or

a temporary one, may control the lower limit of local streams. In arid landscapes, with their intermittent precipitation, valleys, plains, or other low points determine local base level.

• *Construct* a basic drainage basin model and *identify* different types of drainage patterns, with examples.

5. According to Figure 11.5, in which drainage basin are you located? Where are you in relation to the various continental divides?

See Figure 11.5 for a personal response.

6. What is the spatial geomorphic unit of an individual river system? How is it determined on the landscape? Define the key relevant terms used.

Streams are organized into areas or regions called drainage basins. A drainage basin is the spatial geomorphic unit occupied by a river system. A drainage basin is defined by ridges that form drainage divides, (i.e., the ridges are the dividing lines that control into which basin precipitation drains). Drainage divides define watersheds, the catchment areas of the drainage basin. The United States and Canada are divided by several continental divides; these are extensive mountain and highland regions that separate drainage basins, sending flows either to the Pacific, to the Gulf of Mexico and the Atlantic, or to Hudson Bay and the Arctic Ocean.

7. On Figure 11.5, follow the Allegheny–Ohio–Mississippi River systems to the Gulf of Mexico, analyze the pattern of tributaries, and describe the channel. What role do continental divides play in this drainage?

Rainfall in northcentral Pennsylvania generates the streams that flow into the Allegheny River. The Allegheny River then joins with the Monongahela River at Pittsburgh to form the Ohio River. The Ohio flows southwest and connects with the Mississippi River at Cairo, Illinois, and eventually flows on past New Orleans to the Gulf of Mexico. Each contributing tributary adds its discharge and sediment load to the larger river. In our example, sediment weathered and eroded in northcentral Pennsylvania is transported thousands of kilometers and accumulates as the Mississippi delta on the floor of the Gulf of Mexico.

8. Describe drainage patterns. Define the various patterns that commonly appear in nature. What drainage patterns exist in your hometown? Where you attend school?

A drainage basin is the spatial geomorphic unit occupied by a river system. A drainage basin is defined by ridges that form drainage divides, (i.e., the ridges are the dividing lines that control into which basin precipitation drains). Drainage basins are open systems whose inputs include precipitation, the minerals and rocks of the regional geology, and both the uplift and subsidence provided by tectonic activities. System outputs of water and sediment leave through the mouth of the river. Change that occurs in any portion of a drainage basin can affect the entire system as the stream adjusts to carry the appropriate load relative to discharge and velocity. Seven principal drainage patterns are shown in Figure 11.9.

_____

• *Describe* the relation among velocity, depth, width, and discharge and *explain* the various ways that a stream erodes and transports its load.

9. What was the impact of flood discharge on the channel of the San Juan River near Bluff, Utah? Why did these changes take place?

Figure 11.10 shows changes in the San Juan River channel in Utah that occurred during a flood. The increase in discharge increased the velocity and therefore the carrying capacity of the river as the flood progressed. As a result, the river's ability to scour materials from its bed was enhanced. Such scouring represents a powerful clearing action.

10. How does stream discharge complete its erosive work? What are the processes at work on the channel?

Several types of erosional processes are operative. Hydraulic action is the work of turbulence in the water—the eddies of motion. Running water causes friction in the joints of the rocks in a stream channel. A hydraulic squeeze-and-release action works to loosen and lift rocks. As this debris moves along, it mechanically erodes the streambed further through the process of abrasion, with rock particles grinding and carving the streambed.

11.   Differentiate between stream competence and stream capacity.

Competence, which is a stream's ability to move particles of specific size, is a function of stream velocity. The total possible load that a stream can transport is its capacity.

12. How does a stream transport its sediment load? What processes are at work?

Four processes transport eroded materials: solution, suspension, saltation, and traction. Solution refers to the dissolved load of a stream, especially the chemical solution derived from minerals such as limestone or dolomite or from soluble salts. The

suspended load consists of fine particles physically held aloft in the stream, with the finest particles not deposited until the stream velocity slows to near zero. The bed load refers to those coarser materials that are dragged along the bed of the stream by traction or are rolled and bounced along by saltation (from the Latin saltim, which means "by leaps or jumps."

_____

• *Develop* a model of a meandering stream, including point bar, undercut bank, and cutoff, and *explain* the role of stream gradient in these flow characteristics.

13. Describe the flow characteristics of a meandering stream. What is the pattern of the flow in the channel? What are the erosional and depositional features and the typical landforms created?

A meandering channel pattern is common for a stream that slopes gradually, a sinuous form weaving across the landscape. The outer portion of each meandering curve is subject to the greatest erosive action and can be the site of a steep bank called a cut bank (Figure 11.14). On the other hand, the inner portion of a meander receives sediment fill and forms a deposit called a point bar. As meanders develop, these scour-and-fill features gradually work their way downstream. If the load in a stream exceeds the capacity of the stream, sediments accumulate as an aggradation in the stream channel (the opposite of degradation) as the channel builds up through deposition. With excess sediment, a stream becomes a maze of interconnected channels laced with sediments that form a braided pattern.

14. Explain these statements: (a) All streams have a gradient, but not all streams are graded. (b) Graded streams may have ungraded segments.

Every stream has a degree of inclination or gradient, which is the rate of decline in elevation from its headwaters to its mouth, generally forming a concave-shaped slope (Figure 11.17). Theoretically, a stream gradient becomes graded when the load carried by the stream and the landscape through which it flows become mutually adjusted, forming a state of dynamic equilibrium among erosion, transported load, deposition, and the stream's capacity. Attainment of a graded condition does not mean that the stream is at its lowest gradient, but rather that it represents a balance among erosion, transportation, and deposition over time along a specific portion of the stream.

One problem with applying the graded stream concept in an absolute sense, however, is that an individual stream can have both graded and ungraded portions and may have graded sections without having an overall graded slope. In fact, variations and interruptions in a graded profile of equilibrium occur as a rule rather than an exception, making a universally acceptable definition difficult.

15. Why is Niagara Falls an example of a nickpoint? Without human intervention what do you think would eventually take place at Niagara Falls?

At Niagara Falls on the Ontario-New York border, glaciers advanced and receded over the region, exposing resistant rock strata underlain by less-resistant shales. As the less-resistant material continued to weather away, the overlying rock strata collapsed, allowing the falls to erode farther upstream toward Lake Erie. In fact, the falls have retreated more than 11 km (6.8 mi) from the steep face of the Niagara escarpment (long cliff) during the past 12,000 years (see Figure 11.20).

16. Apply these concepts (gradient, graded stream, meandering stream, nickpoint), where appropriate to a stream in your area. Explain and discuss.

Personal response based on text and classroom presentation. Perhaps students could be directed to look at a topo maps for the area.

• *Define* a floodplain and *analyze* the behavior of a stream channel during a flood.

17. Describe the formation of a floodplain. How are natural levees, oxbow lakes, backswamps, and yazoo tributaries produced?

The low-lying area near a stream channel that is subjected to recurrent flooding is a floodplain. It is formed when the river leaves its channel during times of high flow. Thus, when the river channel changes course or when floods occur, the floodplain is inundated with water. When the water recedes, alluvial deposits generally mask the underlying rock. Figure 11.2 illustrates a characteristic floodplain, with the present river channel embedded in the plain's alluvial deposits. The former meander scars form water-filled (often ephemeral) loops on the floodplain called oxbow lakes (known as a *billabong* in Australia, an aboriginal term meaning *dead river* that is only intermittently filled with water).

On either bank of the river are natural levees, which are by-products of flooding. When floodwaters arrive, the river overflows its banks, loses velocity as it spreads out, and drops a portion of its sediment load to form the levees. Larger sand-sized particles drop out first, forming the principal component of the levees, with finer silts and clays deposited farther from the river. Successive floods increase the height of the levees and may even raise the overall elevation of the channel bed so that it is perched above the surrounding floodplain.

Notice on Figure 11.21 an area labeled backswamp and a stream called a yazoo tributary. The natural levees and elevated channel of the river prevent this tributary from joining the main channel, so it flows parallel to the river and through the backswamp area.

18. How is it possible to travel fewer kilometers on the Mississippi River between St. Louis and New Orleans today than 100 years ago? Explain.

As each neck of a meander is cutoff either naturally or by human engineering, the actual distance traveled down the river is decreased. Mark Twain described getting off a riverboat and walking the short distance across the neck while the boat went around the meander, getting some "land time" and a walk along the way.

19. Describe any floodplains near where you live or where you go to college. Have you seen any of the floodplain features discussed in this chapter? If so, which ones?

Personal analysis and response. Check topographic map quadrangles for your area.

• *Differentiate* the several types of river deltas and *detail* each.

20. What is a river delta? What are the various deltaic forms? Give some examples.

The mouth of a river marks the point where the river reaches a base level. Its forward velocity rapidly decelerates as it enters a larger body of standing water, with the reduced velocity causing its transported load to be in excess of its capacity. Coarse sediments drop out first, with finer clays being carried to the extreme end of the deposit. This depositional plain formed at the mouth of a river is called a delta, named after the triangular shape of the Greek letter delta, which was perceived by Herodotus in ancient times to be similar to the shape of the Nile River delta. See the discussion of deltaic forms in the text.

21. How might life in New Orleans change in this century following the events of 2005? What are the floodplain management issues involved? Explain.

Due to the dynamic character of the Mississippi River delta, the main channel of the delta persists in its present location because of much effort and expense directed to maintain an artificial levee system. Compaction and tremendous weight of the sediments in the Mississippi River create isostatic adjustment in Earth's crust. This is causing the entire region of the delta to subside, placing tremendous stress on natural and artificial levees along the lower Mississippi.

The city of New Orleans is almost entirely below river level, with sections of the city below sea level. Severe flooding is a certainty for existing and planned settlements unless further intervention or urban relocation occurs. The building of multiple flood-control structures and extensive reclamation efforts by the U.S. Army Corps of Engineers apparently only delayed the peril, as demonstrated by recent flooding. The fact that four levee breaks caused by Hurricane Katrina's storm surge, winds, and rainfall in 2005 inundated 80% of the city confirms this concern and issue.

An additional problem for the lower Mississippi Valley is the possibility that the river could break from its existing channel and seek a new route to the Gulf of Mexico. The obvious alternative route for the Mississippi is along the Atchafalaya River. If the Mississippi would bypass New Orleans, the threat of flooding would be reduced, yet it would be a financial disaster for New Orleans since the port would silt in and seawater would intrude the freshwater resources.

22. Describe the Ganges River delta. What factors upstream explain its form and pattern? Assess the consequences of settlement on this delta.

The Ganges River delta features an intricate braided pattern of distributaries. Alluvium carried from deforested slopes upstream provides excess sediment that forms the many deltaic islands.

Catastrophic floods continue to be a threat. In Bangladesh, intense monsoonal rains and tropical cyclones in 1988 and 1991 created devastating floods over the country's vast alluvial plain (130,000 km2 or 50,000 mi2). One of the most densely populated countries on Earth, Bangladesh was more than three-fourths covered by floodwaters. Excessive forest harvesting in the upstream portions of the Ganges-Brahmaputra River watersheds increased runoff and added to the severity of the flooding. Over time the increased load carried by the river was deposited in the Bay of Bengal, creating new islands. These islands, barely above sea level, became sites of new settlements and farming villages. When the floodwaters finally did recede, the lack of freshwater—coupled with crop failures, disease, and pestilence—led to famine and the death of tens of thousands. About 30 million people were left homeless and many of the alluvial-formed islands were gone.

23. What is meant by the statement, "the Nile River delta is receding" (caption Figure 11.25)?

The Nile delta is disappearing due to the building of the Aswan Dam and the extensive network of canals that have been built in the delta to augment the natural distributary system. Yet as the river enters the network of canals, flow velocity is reduced, stream competence and capacity are lost, and sediment load is deposited far short of where the delta reaches the Mediterranean Sea. River flows no longer reach the sea! The Nile Delta is receding from the coast at an alarming 50 to 100 m per year. Seawater is intruding farther inland in both surface water and groundwater.

───────────

• *Explain* flood probability estimates and *review* strategies for mitigating flood hazards.

24. Specifically, what is a flood? How are such flows measured and tracked?

A flood is a high water level that overflows the natural (or artificial) banks along any portion of a stream. Understanding flood patterns for a drainage basin is as complex as understanding the weather, for floods and weather are equally variable, and both include a level of unpredictability. The key is to measure streamflow—the height and discharge of a stream. A staff gauge, a pole placed in a stream bank and marked with water heights, is used to measure stream level. With a fully measured cross section, stream level can be used to determine discharge. A stilling well is sited on the stream bank and a gauge is mounted in it to measure stream level. A portable current meter can be used to sample velocity at various locations.

Approximately 11,000 stream gauge stations are used in the United States (an average of over 200 per state). Of these, 7000 have continuous stage and discharge records operated by the U.S. Geological Survey. Many of these stations automatically telemeter data to satellites, from which information is retransmitted to regional centers. Environment Canada's Water Survey of Canada maintains more than 3000 gauging stations.

25. Differentiate between a hydrograph from a natural terrain and one from an urbanized area.

A graph of stream discharge over a time period for a specific place is called a hydrograph. The hydrograph in Figure 11.29 shows the relationship between stream discharge and precipitation input. During dry periods, at low water stages, the flow is described as base flow and is largely maintained by contributions from the local water table. When rainfall occurs in some portion of the watershed, the runoff collects and is concentrated in streams and tributaries. The amount, location, and duration of the rainfall episode determine the peak flow. Also important is the nature of the surface in a watershed; for example, a hydrograph for a specific portion of a stream changes after a forest fire or urbanization of the watershed.

Human activities have enormous impact on water flow in a basin. The effects of urbanization are quite dramatic, both increasing and hastening peak flow as shown in the same figure. In fact, urban areas produce runoff patterns quite similar to those of deserts. The sealed surfaces of the city drastically reduce infiltration and soil moisture recharge, behaving much like the hard, nearly barren surfaces of the desert.

The flood pictured in Figure 11.22 occurred on the Red River during 1997. Many of the concepts discussed in the text under "Floods and River Management" and Focus Study 11.1 "Floodplain Strategies," apply directly to these floods and the experiences of the many cities affected. For historical background on the Midwest floods of 1993 See: Mary F. Myers and Gilbert F. White, "The Challenge of the Mississippi Flood," *Environment*, 35, No. 10, December 1993: 2–9+. This article presents a chronology of "Significant Events in the Development of U.S. Flood Control Policy (1825–1993)."

An extensive portrait by Alan Mairson, "The Great Flood of '93," appears in *National Geographic*, 185, No. 1, January 1994: 42–81. Included are remotely sensed orbital images before and during the flood showing surface water and soil moisture. Also: a summary of two dozen wire-service stories with maps and photographs that appeared in *Storm—The World Weather Magazine*, 1, no. 1, August 1993: 6–16 (1-708-406-8330).

The USGS produced a map of 1993 peak flood stages and discharges as compared to previous maximums experienced in the Mississippi River basin. See: C. Parrett, N.B. Melcher, and R.W. James, Jr., *Flood Discharges in the Upper Mississippi Basin*, Circular 1120-A, Washington, DC: USGS, 1993. A map in the circular shows that discharges were higher than previous maximums at 154 stations.

26. What do you see as the major consideration regarding floodplain management? How would you describe the general attitude of society toward natural hazards and disasters?

Throughout history, civilizations have settled floodplains and deltas, especially since the agricultural revolution that occurred some 8000 years B.P., when the fertility of floodplain soils was discovered. Early villages were generally built away from the area of flooding, or on stream terraces, because the floodplain was the location of intense farming. However, as the era of commerce grew, sites near rivers became important for transportation: port and dock facilities and river bridges to related settlements were built. Also, because water is a basic industrial raw material used for cooling and for diluting and removing wastes, waterside industrial sites became desirable.

Human activities on vulnerable flood-prone lands require planning to reduce or avoid disaster. Essentially, relative to all natural disasters, including floodplains, human societies appear to be unwilling, unable, or incapable of perceiving hazards in a familiar environment. The development and urbanization of the Sacramento River floodplain presented in the discussion a few pages earlier exemplifies these attitudes.

27. What do you think the author of the article, "Settlement Control Beats Flood Control," meant by the title? Explain your answer, using information presented in the chapter.

As suggested in an article, "Settlement Control Beats Flood Control" (Walter Kollmorgen, *Economic Geography* 29, No. 3, July 1953: 215), there are other ways to protect populations than with enormous, expensive, sometimes environmentally disruptive projects. Strictly zoning the floodplain is one approach, but flat, easily developed floodplains near pleasant rivers might be perceived as desirable for housing, and thus weaken political resolve. This strategy would set aside the floodplain for farming or passive recreation, such as riverine parks, golf course, or plant and wildlife sanctuary, or for other uses not hurt by natural floods. Kollmorgen's study concludes that "urban and industrial losses would be largely obviated by set-back levees and zoning and thus cancel the biggest share of the assessed benefits which justify big dams." Focus Study 11.1 takes a closer look at floodplain management.

Overhead Transparencies

12

Wind Processes and Desert Landscapes

Wind is an agent of erosion, transportation, and deposition. Its effectiveness has been the subject of debate; in fact, wind at times was thought to produce major landforms. Presently, wind is regarded as a relatively minor exogenic agent, but it is significant enough to deserve our attention. In this chapter we examine the work of wind, its associated processes, and resulting landforms. In addition, we look at desert landscapes, where water remains the major erosional force but where an overall lack of moisture and stabilizing vegetation allows wind processes to operate. Evidence of this fact are extensive sand seas and sand dunes of infinite variety. Beach and coastal dunes, which form in many climates, also are influenced by wind and are discussed in the next chapter.

Note: A most massive piece of legislation to protect wilderness, in this case desert lands, passed several years ago. The California Desert Protection Act of 1994, passed by Congress and signed by the President. Among other things Death Valley and Joshua Tree were made national parks and a new 1.2 million-acre East Mojave National Preserve was created. The legislation was the kind of bipartisan compromise that could lead to sustainable economic development. In the Preserve hunting and grazing will still be allowed although it will be controlled under the National Park Service instead of the Bureau of Land Management. The BLM will oversee another 3.5 million acres as part of the new law.

Many thought that we would see more of this type of economic-ecological synthesis; yet following the election in 1994, some members of the 104th and 105th sessions of Congress declared war on these set asides of deserts and monument upgrading to parks. At one point Congress set the total operating budget for these new areas at $1!

President Clinton increased the heat when he set aside 1.7 million acres of federal land in Utah, in 1996. Using a 1906 law (Act 16 U.S.C. 431-433, the Antiquities Act of 1906)), the Grand Staircase-Escalante National Monument was established without Congressional involvement. The spectacular scenery is unique in the world. Senator Orrin Hatch (R-Utah) denounced it as the "mother of all land grabs." Subsequently the Utah legislature authorized oil exploration in the new national monument, even though the oil company involved placed low probabilities of 1 in 10 for success. Some believe there are extensive coal reserves beneath the region. These special interests are partially funding grass roots groups to oppose the monument in any form, calling it a land grab and worse.

Meanwhile, as the debate drags along, tourist visits are up 65% over the previous year at the local BLM office. New hotels and motels are going in to the small towns to accommodate the influx. In the past, scenery and tourism has proved much more sustainable of natural systems and local economies than was resource exploitation and the eventual loss of the landscape. See: **http://www.blm.gov/nhp /news/alerts/EscalanteOverview.html**

Outline Headings and Key Terms

The first-, second-, and third-order headings that divide Chapter 12 serve as an outline for your notes and studies. The key terms and concepts that appear **boldface** in the text are listed here under their appropriate heading in ***bold italics***. All these highlighted terms appear in the text glossary. Note the check-off box (❑) so you can mark your progress as you master each concept. Your students have this same outline in their *Student Study Guide*. The ✎ icon indicates that there is an accompanying animation or other resource on the CD.

The outline headings for Chapter 12:

The Work of Wind
- ❑ *eolian*

Eolian Erosion

⊘ **How Wind Moves Sediment**
- ❑ *deflation*
- ❑ *abrasion*

Deflation
- ❑ *desert pavement*
- ❑ *blowout depression*

Abrasion
- ❑ *ventifacts*
- ❑ *yardangs*

Eolian Transportation
- ❑ *surface creep*

Eolian Depositional Landforms

⊘ **Formation of Cross Bedding**
- ❑ *dune*
- ❑ *erg desert*
- ❑ *sand sea*

Dune Movement and Form

Loess Deposits

- ❑ *loess*

Overview of Desert Landscapes

Desert Climates

Desert Fluvial Processes
- ❑ *flash flood*
- ❑ *wash*
- ❑ *playa*

Alluvial Fans
- ❑ *alluvial fan*
- ❑ *bajada*

Desert Landscapes
- ❑ *badland*

Basin and Range Province
- ❑ ***Basin and Range Province***
- ❑ *bolson*

Desertification
- ❑ *desertification*

Summary and Review

News Reports and Focus Study

News Report 12.1: The Dust Bowl

Focus Study 12.1: The Colorado River: A System Out of Balance

---

The URLs related to this chapter of *Elemental Geosystems* can be found at
http://www.prenticehall.com/christopherson

Key Learning Concepts

After reading the chapter and using the Student Study Guide, the student should be able to:

• *Characterize* the unique work accomplished by wind and eolian processes.
• *Describe* eolian erosion including deflation, abrasion, and the resultant landforms.
• *Describe* eolian transportation and *explain* saltation and surface creep.
• *Identify* the major classes of sand dunes and *present* examples within each class.
• *Define* loess deposits, their origins, locations, and landforms.
• *Portray* desert landscapes and *locate* these regions on a world map.

Annotated Chapter Review Questions

• *Characterize* **the unique work accomplished by wind and eolian processes.**

1. Who was Ralph Bagnold? What was his contribution to eolian studies?

A British major, Ralph Bagnold was stationed in Egypt in 1925. Bagnold was an engineering officer who spent much of his time in the deserts west of the Nile, where he measured, sketched, and developed hypotheses about the wind and desert forms. His often-cited work, *The Physics of Blown Sand and Desert Dunes*, was published in 1941 following the completion of wind-tunnel simulations in London. For example, Bagnold studied the ability of wind to transport sand over the surface of a dune. He found that a wind of 50 kmph (30 mph) can move approximately one-half a ton of sand per day over a one-meter-wide section of dune. The graph also demonstrates how rapidly the amount of transported sand increases with wind speed.

I include a quote from Ralph Bagnold, who is credited with the first important arid-land and sand-sea research. While stationed with the British army in Egypt, he explored the western deserts of Egypt. Much of his research time had to be sandwiched in on available leaves from his post. Bagnold took Henry Ford at his word that a Model-T Ford could handle the most difficult terrain. He drove a Model-T all over the desert, carrying sections of chicken wire for areas where support was needed under the wheels. Additional wind-tunnel tests completed key aspects of his research on eolian processes (aeolian is the European spelling).

2. Explain the term *eolian* and its application in this chapter. How would you characterize the ability of the wind to move material?

Wind-eroded, wind-transported, and wind-deposited materials are called eolian (also spelled aeolian; named for Aeolus, the ruler of the winds in Greek mythology). The actual ability of wind to move materials is small compared with that of other transporting agents such as water and ice, because air is so much less dense than these other media.

• *Describe* **eolian transportation, including deflation, abrasion, and the resultant landforms.**

3. Describe erosional processes associated with moving air.

Two principal wind-erosion processes are deflation, the removal and lifting of individual loose particles, and abrasion, the grinding of rock surfaces with a "sandblasting" action by particles captured in the air.

4. Explain deflation and the evolutionary sequence that produces desert pavement.

Deflation literally blows away unconsolidated or noncohesive sediment. After wind deflation and sheetwash do their work on an arid landscape, a desert pavement is formed from the concentrated pebbles and gravels left behind. That pavement, which resembles a cobblestone street, protects underlying sediment from further deflation (Figure 12.2). Desert pavements are so common that many provincial names have been used for them—for example, gibber plain in Australia, gobi in China, and in Africa, lag gravels or serir (or reg desert if some fine particles remain). Water also should be considered in the formation of lag desert plains that wind deflation has modified.

5. How are ventifacts and yardangs formed by the wind?

Rocks exposed to eolian abrasion appear pitted, grooved, or polished, and usually are aerodynamically shaped in a specific direction, according to the flow of airborne particles. Rocks that bear such evidence of eolian erosion are called ventifacts. On a larger scale, deflation and abrasion are capable of streamlining rock structures that are aligned parallel to the most effective wind direction, leaving behind distinctive, elongated ridges called yardangs. These can range from meters to kilometers in length and up to many meters in height (Figure 12.3).

• *Describe* **eolian transportation and *explain* saltation and surface creep.**

6. Differentiate between a dust storm and a sand storm.

Only the finest dust particles travel significant distances, and consequently the finer material suspended in a dust storm is lifted much higher than the coarser particles of a sand storm, which may be lifted only about 2 m (6.5 ft).

7. What is the difference between eolian saltation and fluvial saltation?

The term saltation was used in the river systems Chapter 11 to describe movement of particles along stream beds. The term saltation also is used in eolian processes to describe the wind transport of grains along the ground, grains usually larger than 0.2 mm (0.008 in.). About 80% of wind transport of particles is accomplished by this skipping and bouncing action (Figure 12.5). In comparison with fluvial transport, in which saltation is accomplished by hydraulic lift, eolian saltation is executed by aerodynamic lift, elastic bounce, and impact.

8. Explain the concept of surface creep.

Wind exerts a drag or frictional pull on surface particles. Bagnold studied the relationship between wind velocity and grain size, determining the fluid threshold (minimum wind speed) required for initial movement of grains of various sizes. A slightly lower wind velocity suffices if the particle already has been set into motion by the impact of a saltating grain. Bagnold termed this lesser velocity the impact threshold. Once in motion, particles continue to be transported by lower wind velocities. As you can see from the graph, the impact threshold is less than the fluid threshold.

Notes on the 1991 Persian Gulf War. In addition to the discussion in the text, be sure to mention the importance of intermittent flows of water in the formation of desert pavement, or lag desert deposits. Water delivers materials to help cement the cobbles and pebbles of the surface.

The disruption of desert pavement in the recent Persian Gulf War represents a spatial consequence of war activities in an arid environment, especially when the expenditure of munitions and movement of heavy vehicles are so concentrated in time and area. Several recent references are available, including: Constance Holden's "Kuwait's Unjust Deserts: Damage to Its Desert," *Science* Vol. 251 (8 March 1991): 1175—subtitled "Among the less heralded consequences of the Gulf war could be a massive movement of sand engulfing roads and towns"; also see the survey article by Thomas Y. Canby in the August 1991 *National Geographic* (Vol. 180, No. 2, pp.

2–35). The environment is overlooked in both theater and strategic military planning, yet with modern firepower and mobilization prowess, the environment should be considered.

A recent study completed by Dr. Farouk El-Baz, director of the Center for Remote Sensing at Boston University, reported on the impact of the Persian Gulf War in Kuwait. Using Landsat and *SPOT* images and ground-based confirmation studies he found that 5446 km^2 of Kuwait's total land-surface area suffered war related damage—that's 30.6% of Kuwait! Impacts were classified as land mines and post-war cleanup of mines and unexploded ordinance, land blackened by oil and soot, destroyed desert vegetation, land altered by the construction of berms, oil lakes, oil-well fire fighting alteration of the land, and vehicle tracks. The overall destruction produced by these impacts destroyed the desert pavement. Freed fine-grained materials were then available for deflation and transport, covering farms and roads, and encroaching on cities. Groundwater resources are being watched closely for oil contamination as it percolates to the water table. Dr. El-Baz presented these results at the October 1993 meeting of the Geological Society of America (GSA) in Boston.

These are issues geographers are equipped to handle. With adequate data inputs and a GIS-type work up on the potential impacts of the war. We wonder what weight such a GIS approach to the environment would have in encouraging the economic-sanction alternative to going to war. Also see: "Burning Questions" and "Why are Data from Kuwait Being Withheld?" in *Scientific American* 265, No. 1 (July 1991); and, "Up in Flames" and "U.S. Gags Discussion of War's Environmental Effects," in *Scientific American* 264, No. 5 (May 1991): 17–24; and, Sir Frederick Warner's "The Environmental Consequences of the Gulf War," *ENVIRONMENT* 33, No. 5 (June 1991): 7–9, 25–26.

As an example of a proactive statement about strategic war, please consult Association of American Geographers' "Statement of the Association of American Geographers on Nuclear War." Adopted by AAG, 6 May 1986, Minneapolis.

• *Identify* the major classes of sand dunes and *present* examples within each class.

9. What is the difference between an erg and a reg desert? Which type is a sand sea? Are all deserts covered by sand? Explain.

A common assumption is that most deserts are covered by sand. Instead, desert pavements predominate across most sub-tropical arid landscapes; only about 10% of desert areas are covered with sand.

Sand grains generally are deposited as transient ridges or hills called dunes. A dune is a wind-sculpted accumulation of sand. An extensive area of dunes, such as that found in North Africa, is characteristic of an erg desert, which means sand sea. Most desert landscapes are not covered with sand but are desert pavements, which are so common that many provincial names have been used for them—for example, gibber plain in Australia, gobi in China, and in Africa, lag gravels or serir or reg desert.

We recommend that you obtain a copy of McKee, Edwin D. *A Study of Global Sand Seas*, U.S. Geological Survey Professional Paper 1052. Washington: Government Printing Office, 1979. You can get this through any one of the USGS regional bookstores. This oversized book is an excellent comprehensive treatment of erg desert landscapes and forms. The dune illustrations in Figure 12.10 were adapted from this publication.

10. What are the three classes of dune forms? Describe the basic types of dunes within each class. What do you think is the major shaping force for sand dunes?

We can simplify dune forms into three classes—crescentic, linear, and star dunes. See Figure 12.10 for classes of dunes, their descriptions, and an illustration of each major type.

11. Which form of dune is the mountain giant of the desert? What are the characteristic wind patterns that produce such dunes?

Star dunes are the mountain giants of the sandy desert. They form in response to complicated, changing wind patterns and have multiple slipfaces. They are rosette-shaped or pinwheel-shaped, with several radiating arms rising and joining to form a common central peak or crest. The best examples of star dunes are in the Sahara, where they approach 200 m (650 ft) in height (Figure 12.11).

• *Define* loess deposits, their origins, locations, and landforms.

12. How are loess materials generated? What form do they assume when deposited?

Pleistocene glaciers advanced and retreated in many parts of the world, leaving behind large glacial outwash deposits of fine-grained clays and silts (<0.06 mm or 0.0023 in.). These materials were blown great distances by the wind and redeposited in unstratified, homogeneous deposits named loess. Loess deposits form some complex weathered badlands and some good agricultural land.

13. Name a few examples of significant loess deposits on Earth.

See Figure 12.14. In Europe and North America, loess is thought to be derived mainly from glacial and periglacial sources. The vast deposits of loess in China, covering more than 300,000 km² (115,800 mi²), are thought to be derived from desert rather than glacial sources. Accumulations in the Loess Plateau of China exceed 300 m (984 ft) of thickness.

• *Portray* desert landscapes and *locate* these regions on a world map.

14. Characterize desert energy and water balance regimes. What are the significant patterns of occurrence for arid landscapes in the world?

The daily surface energy balance for El Mirage, California, presented in Figure 3.13, highlights the high sensible heat conditions and intense ground heating in the desert. Such areas receive a high input of insolation through clear daylight skies and experience high radiative heat losses at night. A typical desert water balance is given in Figure 7.10b, for Phoenix, Arizona, showing high potential evapotranspiration demand, low precipitation supply, and prolonged summer deficits. Fluvial processes in the desert generally are dominated by intermittent running water, with hard, sparsely vegetated desert pavement yielding high runoff during rainstorms. The spatial distribution of these dry lands is related to several factors: subtropical high-pressure cells between 15° and 35° N and S, rain shadows to the lee of mountain ranges, and great distance from moisture-bearing air masses. Figure 12.15 portrays this distribution according to the climate classification used in this text. The photos around the map illustrate well that all deserts do not look alike.

15. How would you describe the water budget of the Colorado River? What was the basis for agreements regarding distribution of the river? Why has thinking about the river's discharge been so optimistic?

Exotic stream flows are highly variable, and the Colorado is no exception. In 1917, the discharge measured at Lees Ferry totaled 24 million acre-feet (maf), whereas in 1934 it reached only 5.03 maf. In 1977, the discharge dropped to 5.02 but rose to an all-time high of 24.5 maf in 1984. In addition to this variability, approximately 70% of the year-to-year discharge occur between April and July.

The average flows between 1906 and 1930 were almost 18 maf a year, with averages dropping to 14.2 maf during the 1930–1998 period. As a planning basis for the Colorado River Compact, the government used average river discharges from 1914 up to the treaty signing in 1923, an exceptionally high 18.8 maf. That amount was perceived as more than enough for the upper and lower basins each to receive 7.5 maf and, later, enough for Mexico to receive 1.5 maf in the 1944 Mexican Water Treaty.

We hope discussion and dialog result from this Focus Study, since the budget presented in Table 12.1.1 demonstrates a river over budgeted by some 5.8 million acre-feet per year. Present understanding of Colorado River flows seems to be based on the exceptionally high river discharges that occurred before the Colorado River Compact was signed in 1922.

The floods generated by the El Niño event of 1982–83 produced near-record discharges in the river. Federal regulators, under the leadership of the Secretary of the Interior, apparently saw an opportunity to have all the reservoirs in the system full for the first time in the history of reclamation activities on the river. In essence this meant that the government would invert the reclamation law that demands flood control as the prime function of these multi-purpose facilities. Instead, water for irrigation and hydroelectric interests were placed first. The resultant flood described in the Focus Study occurred—a human-caused flood along the most regulated river in the world!

A few dry years in a row will certainly stress the river's water budget, especially with the operation of the new Central Arizona Project (CAP) and increasing demands by the upper basin states for their allotted shares. Perhaps the most telling fact of irrigating the desert occurs in the Wellton-Mohawk irrigation area (east of Yuma, Arizona). It would have been cheaper for the government to buy the land at a fair market price and take it out of production than it cost to subsidize the irrigation and to construct a desalinization plant for the runoff drainage from the irrigated farmland.

16. Describe a desert bolson from crest to crest. Draw a simple sketch with the components of the landscape labeled.

Figure 12.23c illustrates a typical bolson, which is a slope-and-basin area between the crests of two adjacent ridges in a dry region. Basin-and-range relief is abrupt, and rock structures are angular and rugged. As the ranges erode, the transported materials accumulate to great depths in the basins, gradually producing extensive desert plains.

17. Where is the Basin and Range Province? Briefly describe its appearance and character.

The basins average 1220–1525 m (4000–5000 ft) above sea level, with mountain crests rising higher by some 915–1525 m (3000–5000 ft). Death Valley is an extreme, being the lowest of these basins, with an elevation of –86 m (–282 ft). However, to the west of the valley, the Panamint Range rises to 3368 m (11,050 ft) at Telescope Peak—over 3 vertical kilometers (2 mi) of desert relief!

18. What is meant by desertification? Using the map in Figure 12.24, and the text description, locate several of the affected regions of desert expansion.

We are witnessing an unwanted expansion of the Earth's desert lands in a process known as desertification. This now is a worldwide phenomenon along the margins of semiarid and arid lands. Desertification is due principally to poor agricultural practices (overgrazing and agricultural activities that abuse soil structure and fertility), improper soil-moisture management, erosion and salinization, deforestation, and the ongoing global climatic change which is shifting temperature and precipitation patterns. The United Nations estimates that degraded lands have covered some 800 million hectares (2 billion acres) since 1930; many millions of additional hectares are added each year. An immediate need is to improve the data base for a more accurate accounting of the problem and a better understanding of what is occurring.

Figure 12.22, p. 393, is drawn from a map prepared for a U.N. Conference on Desertification. Desertification areas are ranked: A moderate hazard area has an average 10%–25% drop in agricultural productivity; a high hazard area has a 25%–50% drop; and a very high hazard area has more than a 50% decrease. Because human activities and economies, especially unwise grazing practices, appear to be the major cause of desertification, actions to slow the process are readily available. The severity of this problem is magnified by the poverty in many of the affected regions.

Overhead Transparencies

230.	12.2 a, b	Desert pavement formation; desert pavement photo
231.	12.5 a, b	Eolian saltation and surface creep (top); moving sand (bottom)
232.	12.10 (left)	Major dune forms (left page) part 1
233.	12.10 (right)	Major dune forms (right page) part 2
234.	12.15	The world's arid lands map
235.	12.17	Alluvial fan: map and photo
236.	12.21 a – e	Basin and Range Province of the western U. S.
237.	12.22	Desertification hazard, world map
238.	F.S. 12.1.1	The Colorado River Basin

13

The Oceans, Coastal Processes, and Landforms

Coastal regions are unique and dynamic environments. Most of Earth's coastlines are relatively new and are the setting for continuous change. The land, ocean, and atmosphere interact to produce waves, tides, erosional features, and depositional features along the continental margins. The interaction of vast oceanic and atmospheric masses is dramatic along a shoreline. At times, the ocean attacks the coast in a stormy rage of erosive power; at other times, the moist sea breeze, salty mist, and repetitive motion of the water are gentle and calming.

You will find coastal processes in this chapter organized in a system of specific inputs, actions, and outputs. This should permit you to conduct the students through a logical progression of subjects. We conclude with a look at the considerable human impact on coastal environments. As global change affects Earth's coastlines the need for more considerate planning of these potentially hazardous zones becomes more critical. Issues of sea-level rise and coastal inundation, coral bleaching, a tendency toward more devastating tropical storms, loss of beaches and barrier islands, and pollution, will increase interest and awareness.

Outline Headings and Key Terms

The first-, second-, and third-order headings that divide Chapter 13 serve as an outline. The key terms and concepts that appear **boldface** in the text are listed here under their appropriate heading in ***bold italics***. All these highlighted terms appear in the text glossary. Note the check-off box (❑) so you can mark class progress. Your students have this same outline in their Student Study Guide. The ✎ icon indicates that there is an accompanying animation or other resource on the CD.

The outline headings for Chapter 13:
Global Oceans and Seas
　　Chemical Composition of Seawater
　　　　❑ ***salinity***
　　Ocean Chemistry
　　Average Salinity: 35‰
　　　　❑ ***brine***
　　　　❑ ***brackish***
　　Physical Structure of the Ocean
Coastal System Components
　　The Coastal Environment and Sea Level
　　　　❑ ***littoral zone***
　　　　❑ ***mean sea level***
Coastal System Actions
✎ **Monthly Tidal Cycle**
✎ **Wave Motion/Wave Refraction**
✎ **Beach Drift, Coastal Erosion**
　　Tides
　　　　❑ ***tides***
　　Causes of Tides
　　Spring and Neap Tides
　　Tidal Power
　　Waves
　　　　❑ ***waves***
　　　　❑ ***swells***
　　　　❑ ***breaker***
　　Wave Refraction
　　　　❑ ***wave refraction***
　　　　❑ ***longshore current***
　　Tsunami, or Seismic Sea Wave
　　　　❑ ***tsunami***
Coastal System Outputs
✎ **Coastal Stabilization Structures**
✎ **Hurricane Isabel 9/6-9/19/2004**
✎ **Hurricane Georges satellite loop**
　　Erosional Coastal Processes and Landforms
　　　　❑ ***wave-cut platform***
　　Depositional Coastal Processes and Landforms

The URLs related to this chapter of *Elemental Geosystems* can be found at
http://www.prenticehall.com/christopherson

Key Learning Concepts

After reading the chapter and using the Student Study Guide, the student should be able to:

• *Describe* the chemical composition of seawater and the physical structure of the ocean.
• *Identify* the components of the coastal environment and *list* the physical inputs to the coastal system, including tides and mean sea level.
• *Describe* wave motion at sea and near shore and *explain* coastal straightening as a product of wave refraction.
• *Identify* characteristic coastal erosional and depositional landforms.
• *Describe* barrier islands and their hazards as they relate to human settlement.
• *Assess* living coastal environments: corals, wetlands, salt marshes, and mangroves.

• *Construct* an environmentally sensitive model for settlement and land use along the coast.

Annotated Chapter Review Questions

• *Describe* the chemical composition of seawater and the physical structure of the ocean.

1. Describe the salinity of seawater: its composition, amount, and distribution.

Water acts as a solvent, dissolving at least 57 of the elements found in nature. In fact, most natural elements and the compounds they form are found in the seas as dissolved solids, or solutes. Thus, seawater is a solution, and the concentration of dissolved solids is called salinity. Seven elements comprise more than 99% of the dissolved solids in seawater: chlorine (Cl), sodium (Na), magnesium (Mg), sulfur (S), calcium (Ca), potassium (K), and bromine (Br). Seawater also contains dissolved gases (such as carbon dioxide, nitrogen, and oxygen), solid and dissolved organic matter, and a multitude of trace elements. Salinity worldwide normally varies between 34‰ and 37‰; variations are attributable to atmospheric conditions above the water and to the quantity of freshwater inflows. The term brine is applied to water that exceeds the average of 35‰ salinity, whereas brackish applies to water that is less than 35‰.

The section presents at least four ways of expressing the average worldwide value of 35‰. Water as the universal solvent has dissolved an enormous quantity of materials, which is expressed as its salinity. Figure 13.2 shows the variation of salinity with latitude. The following table lists the concentration of some of the elements in a cubic kilometer (and cubic mile) of seawater.

Concentration of elements in km^3 of seawater.

Element	Tons / km^3	Tons / mi^3
Hydrogen(H) + Oxygen (O) = H_2O	991,000,000	4,546,000,000
Chlorine(C)	19,600,000	89,500,000
Sodium (Na)	10,900,000	49,500,000
Magnesium (Mg)	1,400,000	6,125,000
Sulfur (S)	920,000	4,240,000
Calcium (Ca)	420,000	1,880,000
Potassium (K)	390,000	1,790,000
Bromine (Br)	67,000	306,000
Carbon (C)	29,000	132,000
Strontium (St)	8,300	38,000

2. Review the table of ocean information in Figure 13.1. Locate each of the four oceans and the Southern Ocean on a map.

See Figure 13.1 and table—an action item.

3. What are the three general zones relative to physical structure within the ocean? Characterize each by temperature, salinity, dissolved oxygen, and dissolved carbon dioxide.

The ocean's surface layer is warmed by the Sun and is wind-driven. Variations in water temperature and solutes are blended rapidly in a mixing zone that represents only 2% of the oceanic mass. Below this is the thermocline transition zone, a region of strong temperature gradient that lacks the motion of the surface. Friction dampens the effect of surface currents, with colder water temperatures at the lower margin tending to inhibit any convective movements. Starting at a depth of 1–1.5 km (0.62-0.93 mi) and going down to the bottom, temperature and salinity values are quite uniform. Temperatures in this deep cold zone are near 0°C (32°F); but, due to its salinity, seawater freezes at about –2°C (28.4°F). The coldest water is at the bottom except near the poles, where cold water may be near or at the surface. Refer to graphs on right side of Figure 13.2.

_____

• *Identify* **the components of the coastal environment and** *list* **the physical inputs to the coastal system, including tides and mean sea level.**

4. What are the key terms used to describe the coastal environment?

See Figure 13.3. The coastal environment is called the littoral zone. (Littoral comes from the Latin word for shore.) The littoral zone spans both land and water. Landward, it extends to the highest water line that occurs on shore during a storm. Seaward, it extends to the point at which storm waves can no longer move sediments on the seafloor (usually at depths of approximately 60 m or 200 ft). The specific contact line between the sea and the land is the shoreline, and adjacent land is considered the coast.

5. Define mean sea level. How is this value determined? Is it constant or variable around the world? Explain.

Mean sea level is a calculated value based on average tidal levels recorded hourly at a given site over a period of at least 19 years, which is one full lunar tidal cycle. Mean sea level varies spatially from place to place because of ocean currents and waves, tidal variations, air temperature and pressure differences, and ocean temperature variations.

6. What interacting forces generate the pattern of tides?

Earth's orientation to the Sun and the Moon and the reasons for the seasons are discussed in Chapters 2 and 3. The same astronomical relationships produce the pattern of tides, the complex daily oscillations in sea level that are experienced to varying degrees around the world. Tides also are influenced by the size, depth, and topography of ocean basins, by latitude, and by shoreline configuration. Tides are produced by the gravitational pull exerted on Earth by both the Sun and the Moon. Although the Sun's influence is only about half that of the Moon (46%) because of the Sun's greater distance from Earth, it is still a significant force. Figure 13.5 illustrates the relationship among the Moon, the Sun, and Earth and the generation of variable tidal bulges on opposite sides of the planet.

7. What characteristic tides are expected during a new Moon or a full Moon? During the first-quarter and third-quarter phases of the Moon? What is meant by a flood tide? An ebb tide?

Figure 13.5a shows the Moon and the Sun in conjunction (lined up with Earth–new Moon or full Moon), a position in which their gravitational forces add together. The combined gravitational effect is strongest in the conjunction alignment and results in the greatest tidal range between high and low tides, known as spring tides. (Relative to tides, spring means to "spring forth"; it has no relation to the season of the year.) Figure 13.5b shows the other alignment that gives rise to spring tides, when the Moon and Sun are at opposition. When the Moon and the Sun are neither in conjunction nor in opposition, but are more or less in the positions shown in c and d (first- and third-quarter phases), their gravitational influences are offset and counteract each other somewhat, producing a lesser tidal range known as neap tide. Every 24 hours and 50 minutes, any given point on Earth rotates through two bulges as a direct result of this rotational positioning. Thus, every day most coastal locations experience two high (rising) tides known as flood tides, and two low (falling) tides known as ebb tides. The difference between consecutive high and low tides is considered the tidal range.

8. Is tidal power being used anywhere to generate electricity? Explain briefly how such a plant would utilize the tides to produce electricity. Are there any sites in North America? Where are they?

The fact that sea level changes daily with the tides suggests an opportunity that these could be harnessed to produce electricity. The bay or estuary under consideration must have a narrow entrance suitable for the construction of a dam with gates and locks, and it must experience a tidal range of flood and ebb tides large enough to turn turbines, at least a 5 m

(16 ft) range. Tidal power generation is possible at about 30 locations in the world, although, at present, only three of them are actually producing electricity: a 4 megawatt station in Russia at Kislaya-Guba Bay, on the White Sea, since 1968; and a facility in the Rance River estuary on the Brittany coast of France since 1967.

The third site is on the Bay of Fundy. According to the Canadian government, the present cost of tidal power at ideal sites is economically competitive with that of fossil fuels, although certain environmental concerns must be addressed. At one of several favorable sites on the Bay of Fundy, the Annapolis Tidal Generating Station was built in 1984. Nova Scotia Power Incorporated operates the 20 megawatt plant (Figure 13.6d).

• *Describe* wave motion at sea and near shore and *explain* coastal straightening as a product of wave refraction.

9. What is a wave? How are waves generated, and how do they travel across the ocean? Does the water travel with the wave? Discuss the process of wave formation and transmission.

Undulations of ocean water called waves travel in wave trains, or groups of waves. Storms around the world generate large groups of wave trains. A stormy area at sea is the generating region for these large waves, which radiate outward from their formation center. As a result, the ocean is crisscrossed with intricate patterns of waves traveling in all directions. The waves seen along a coast may be the product of a storm center thousands of kilometers away. Water within such a wave is not really migrating but is transferring energy through the water in simple cyclic undulations, which form waves of transition. In a breaker, the orbital motion of transition gives way to waves of translation, in which both energy and water move forward toward shore as water cascades down from the wave crest (Figure 13.7).

10. Describe the refraction process that occurs when waves reach an irregular coastline. How is the coastline straightened?

Generally, wave action is a process that results in coastal straightening. As waves approach an irregular coast, they bend and focus around headlands, or protruding landforms generally composed of more resistant rocks (Figure 13.8). Thus, headlands represent a specific point of wave attack along a coastline. Waves tend to disperse their energy in coves and bays on either side of the headlands. This wave refraction (wave bending) along a coastline redistributes wave energy so

that different sections of the coastline are subjected to variations in erosion potential.

11. Define the components of beach drift and the longshore current and longshore drift.

Particles on the beach are moved along as beach drift, or littoral drift, shifting back and forth between water and land in the effective wind and wave direction. These dislodged materials are available for transport and eventual deposition in coves and inlets and can represent a significant volume. The longshore current transports beach drift. A longshore current generates only in the surf zone and works in combination with wave action to transport large amounts of sand, gravel, sediment and debris along the shore as longshore drift.

12. Explain how a seismic sea wave attains such tremendous velocities. Why is it given a Japanese name?

Tsunamis, often incorrectly referred to as tidal waves, are formed by sudden and sharp motions in the seafloor, such as those caused by earthquakes, submarine landslides, or eruptions of undersea volcanoes. Usually, a solitary wave of great wavelength, or sometimes a small group of two or three long waves, is stimulated. These waves generally exceed 100 km (60 mi) in wavelength but are only a meter or so in height. Because of their great wavelength, tsunamis are affected by the topography of the deep-ocean floor and are refracted by rises and ridges. They travel at great speeds in deep-ocean water—velocities of 600 to 800 kmph (375–500 mph) are not uncommon–but often pass unnoticed because their great length makes the slow rise and fall of water hard to observe. Throughout history Japan has suffered greatly from these seismic waves. Tsunami is Japanese for "harbor wave," named for its devastating effect in harbors.

• *Identify* characteristic coastal erosional and depositional landforms.

13. What is meant by an erosional coast? What are the expected features of such a coast?

The active margins of the Pacific along the North and South American continents are characteristic coastlines affected by erosional landform processes. Erosional coastlines tend to be rugged, of high relief, and tectonically active, as expected from their association with the leading edge of a drifting lithospheric plate. Sea cliffs are formed along a coastline by the undercutting action of the sea. As indentations are produced at water level, such a cliff becomes notched, leading to subsequent collapse and retreat of the cliff. Other erosional forms evolve along cliff-dominated coastlines, including sea caves, sea arches, and sea stacks. As erosion continues, arches may collapse, leaving isolated stacks out in the water. The coasts of southern England and Oregon are prime examples of such erosional landscapes. See Figure 13.10 and inset photos.

14. What is a depositional coast? What are the expected features of such a coast?

Depositional coasts generally are located near onshore plains of gentle relief, where sediments are available from many sources. Such is the case with the Atlantic and Gulf coastal plains of the United States, which lie along the relatively passive, trailing edge of the North American lithospheric plate. A spit consists of sand deposited in a long ridge extending out from a coast; it partially crosses and blocks the mouth of a bay. Classic examples include Sandy Hook, New Jersey, and Cape Cod, Massachusetts. The spit becomes a bay barrier if it completely cuts the bay off from the ocean and forms an inland lagoon. Spits and bars are made up of materials that have been eroded and transported by littoral drift; for much sand to accumulate, offshore currents must be weak. A tombolo occurs when sand deposits connect the shoreline with an offshore island or sea stack. See Figures 13.11c and 13.1.1b.

15. How do people attempt to modify littoral drift? What strategies are used? What are the positive and negative impacts of these actions?

Particles on the beach are moved along as beach drift, shifting back and forth between water and land in the effective wind and wave direction with each swash and backwash of surf. Individual sediment grains trace arched paths along the beach. You have perhaps stood on a beach and heard the sound of myriad grains and seawater especially in the backwash of surf. These dislodged materials are available for transport and eventual deposition in coves and inlets and can represent a significant volume.

Figure 13.12 illustrates several of the common approaches: jetties to block material from harbor entrances, groins to slow drift action along the coast, and a breakwater to create a zone of still water near the coastline. However, interrupting the coastal drift that is the natural replenishment for beaches may lead to unwanted changes in sand distribution downcurrent. In addition, enormous energy and materials must be committed to counteract the enormous and relentless energy that nature invests along the coast.

16. Describe a beach—its form, composition, function, and evolution.

A beach is that place along a coast where sediment is in motion. Material from the land temporarily resides there while it is in active transit along the shore. The beach zone ranges, on average, from 5 m (16 ft) above high tide to 10 m (33 ft) below low tide, although specific definition varies greatly along individual shorelines. Beaches are dominated by quartz (SiO_2) because it is the most abundant mineral on Earth, resists weathering, and therefore remains after other minerals are removed. In volcanic areas, beaches are derived from wave-processed lava. A beach acts to stabilize a shoreline by absorbing wave energy, as is evident by the amount of material that is in almost constant motion.

Some beaches are continuous and stable. Others cycle seasonally; they accumulate during the summer only to be removed by winter storm waves, and are again replaced the following summer. During quiet, low-energy periods, both backshore and foreshore areas experience net accumulations of sand and are well defined. However, a storm brings increased wave action and high-energy conditions to the beach, usually reducing its profile to a sloping foreshore only. Sediments from the backshore area do not disappear but may be moved by wave action to the nearshore (offshore) area, only to be returned by a later low-energy surf.

17. What success has Miami had with beach replenishment? Is it a practical strategy?

See News Report 13.3. In Miami Beach, an 8-year replenishment cycle is maintained. During Hurricane Andrew, in 1992, the replenished Miami Beach is thought to have prevented millions of dollars in shoreline structural damage. The cost of beach replenishment following the 2004 and 2005 hurricane onslaught and sand loss is still undetermined. A further disadvantage is that unforeseen environmental impacts may accompany the addition of new and foreign sand types to a beach.

Beach nourishment refers to the artificial placement of sand along a beach. Through such efforts, a beach that normally experiences a net loss of sediment will instead show a net gain. In contrast to hard structures this hauling of sand to replenish a beach is considered "soft" shoreline protection.

Enormous energy and material must be committed to counteract the relentless energy that nature invests along the coast. Years of human effort and expense can be erased by a single storm, such as when offshore islands were completely eliminated by Hurricane Camille along the Gulf Coast in 1969. In Florida, the city of Miami and surrounding Dade County have spent almost $70 million since the 1970s rebuilding their beaches, and needless to say, the effort is continuous. To maintain a 200 m-wide beach (660 ft), net sand loss per year is determined and a schedule of needed replenishment is set.

Unforeseen environmental impact may accompany the addition of new and foreign sand types to a beach. If the new sands do not match the existing varieties, untold disruption of coastal marine life is possible. The U.S. Army Corps of Engineers, which operates the Miami replenishment program, is running out of "borrowing areas" for sand that matches the natural sand of the beach. A proposal to haul a different type of sand from the Bahamas is being studied as to possible environmental consequences.

_____

• *Describe* **barrier islands and their hazards as they relate to human settlement.**

18. On the basis of the information in the text and any other sources at your disposal, do you think barrier islands and beaches should be used for development? If so, under what conditions? If not, why not?

(The first portion of this question asks the student for an opinion as to hazard perception.) Offshore sandbars gradually migrate toward shore as the sea level rises. Because many barrier beaches evidence this landward migration today, they are an unwise choice for a home site or commercial building. Nonetheless, they are a common choice, even though they take the brunt of storm energy and actually act as protection for the mainland. The hazard represented by the settlement of barrier islands was made graphically clear when Hurricane Hugo (1989) assaulted South Carolina. Beachfront houses, barrier beach developments, and millions of tons of sand were swept away; up to 95% of the single-family homes in Garden City alone were destroyed.

19. After Hurricane Hazel destroyed the Grand Strand off South Carolina in 1954, settlements were rebuilt, only to be hit by Hurricane Hugo 35 years later, in 1989. Why do these recurring events happen to human populations? After the storm events along the Gulf coast in 2005, what do you think the future holds —relocate, rebuild, abandon?

Some homes that had escaped the last major storm, Hurricane Hazel in 1954, were lost to Hugo's peak winds of 209 kmh (130 mph). In comparable dollars, Hugo caused nearly $4 billion in damage, compared with Hazel's $1 billion. Hurricanes Georges and Katrina have had devastating effects on the Chandeleur Islands (Figure 13.14c). The increased damage partially resulted from expanded construction during the intervening years between the two storms. Due to increased development and real estate appreciation, each future storm can be expected to cause ever-increasing capital losses. The inability, unwillingness, or incapacity of humans to perceive

hazards in a familiar environment is well-illustrated. This lack of perception allows the political-business sector to proceed with development and rezoning or non-zoning of high-risk areas. There are many examples of an informed citizenry blocking such efforts by business. The irony is that the conservative atmosphere that seems to prevail across the country relative to taxes does not in some way translate to proper hazard zoning. After all, the taxpayer pays for the reconstruction through disaster benefits and assistance.

Personal response, although a strictly rational response would be abandonment.

_____

• *Assess* **living coastal environments: corals, wetlands, salt marshes, and mangroves.**

20. How are corals able to construct reefs and islands?

A coral is a simple marine animal with a cylindrical, saclike body; it is related to other marine invertebrates, such as anemones and jellyfish. Corals secrete calcium carbonate ($CaCO_3$) from the lower half of their bodies, forming a hard external skeleton. Although both solitary and colonial corals exist, it is the colonial forms that produce enormous structures, varying from treelike and branching forms to round and flat shapes. Through many generations, live corals near the ocean's surface build on the foundation of older corals below, which in turn may rest upon a volcanic seamount or some other submarine feature built up from the ocean floor. An organically derived sedimentary formation of coral rock is called a reef and can assume one of several distinctive shapes, principally a fringing reef, a barrier reef, or an atoll.

21. Describe a trend in corals that is troubling scientists, and discuss some possible causes.

Scientists at the University of Puerto Rico and NOAA are tracking an unprecedented bleaching and dying-off of corals worldwide. The Caribbean, Australia, Japan, Indonesia, Kenya, Florida, Texas, and Hawaii are experiencing this phenomenon. By the end of 2000, approximately 30% of reefs were lost. The bleaching is due to a loss of colorful algae from within and upon the coral itself. Normally colorful corals have turned stark white as the host coral expels nutrient-supplying algae. Exactly why the coral ejects its living partner is unknown. Possibilities include local pollution, disease, sedimentation, and changes in salinity.

Another probable cause is the 1 to 2C° (1.8–3.6F°) warming of sea-surface temperatures, as stimulated by greenhouse warming of the atmosphere. During the 1982–1983 ENSO (Chapter 6) areas of the

Pacific Ocean were warmer than normal and widespread coral bleaching occurred. Coral bleaching worldwide is continuing as average ocean temperatures climb higher. Further bleaching through the decade will affect most of Earth's coral-dominated reefs and may be an indicator of serious and enduring environmental trauma. If present trends continue most of the living corals on Earth could perish within a decade.

22. Why are coastal wetlands poleward of 30° N and S latitude different from those that are equatorward? Describe the differences.

In terms of wetland distribution, salt marshes tend to form north of the 30th parallel, whereas mangrove swamps form equatorward of that point. This is dictated by the occurrence of freezing conditions, which control the survival of mangrove seedlings. Roughly the same latitudinal limits apply in the Southern Hemisphere. Salt marshes usually form in estuaries and behind barrier beaches and sand spits. An accumulation of mud produces a site for the growth of halophytic (salt-tolerant) plants. Plant growth then traps additional alluvial sediments and adds to the salt marsh area. Sediment accumulation on tropical coastlines provides the site for growth of mangrove trees, shrubs, and other small trees. The adventurous prop roots of the mangrove are constantly finding new anchorages and are visible above the water line but reach below the water surface, providing a habitat for a multitude of specialized life forms. Mangrove swamps often secure and fix enough material to form islands.

_____

• *Construct* an environmentally sensitive model for settlement and land use along the coast.

23. Describe the condition of the Chandeleur Islands in the Gulf of Mexico before and after Hurricane Georges in 1998 and Hurricane Katrina in 2005 (Figure 13.14). What does this tell us about the fate of other barrier islands?

The barrier islands off the Louisiana shore are disappearing at rates approaching 20 m (65 ft) per year. Hurricanes have taken their toll on these barrier islands. Also, they are affected by subsidence through compaction of Mississippi delta sediments and a changing sea level that is rising at 1 cm (0.4 in.) per year in the region. Hurricane Andrew eroded sand from 70 percent of barrier islands in 1992. In 1998, Hurricane Georges destroyed large tracts of the Chandeleur Islands (30 to 40 km, 19 to 25 mi, from the Louisiana and Mississippi Gulf Coast), leaving the mainland more exposed (see Figure 13.14). The regional office of the USGS predicts that in a few decades the barrier islands may be gone. Louisiana's increasingly exposed wetlands are disappearing at rates

of 65 km² (25 mi²) (up from 40 km² (15 mi²) per year previously). Hurricane Katrina (2005) alone removed this amount and swept away most of the remaining Chandeleur Islands. These failing barrier islands offer important protection to the Mississippi delta.

24. What type of environmental analysis is needed for rational development and growth in a region like the New Jersey shore? Examine, analyze, and explain the photographs in Focus Study 13.1, Figure 13.1.1. Evaluate South Carolina's approach to coastal hazards and protection.

The key to environmental planning and zoning is to allocate responsibility and cost in the event of a disaster. Appropriate planning assesses the level of hazard that exists on to avoid recurrent disasters. Such assessments address these ecosystems as an integrated unit in order to assess the damage any development would have upon the entire system.

See the analysis completed by Ian McHarg as presented in Focus Study 13.1. Prior to an intensive study completed in 1962, no analysis had been completed outside academic circles. Thus, what is common knowledge to botanists, biologists, ecologists, and geographers in the classroom and laboratory still has not filtered through to the general planning and political processes. As a result, on the New Jersey shore and along much of the Atlantic and Gulf coasts, improper development of the fragile coastal zone led to extensive destruction during the storms of March 1962. The New Jersey shore provides an excellent example of how proper understanding of a coastal environment could have avoided problems encountered during major storms and from human development and settlement.

South Carolina enacted their Beach Management Act of 1988 (modified 1990) to apply some of the principles McHarg described in his analysis of the New Jersey shore. In its first two years, more than 70 lawsuits protested the act as an invalid seizure of private property without compensation. Vulnerable coastal areas are protected from new construction or rebuilding. Now guided by law are structure size, replacement limits for damaged structures, and placement of structures on lots, although liberal interpretation is allowed. The Act sets standards for different coastal forms: beach and dunes, eroding shorelines, or a hypothetical baseline along a coast that lacks dunes or has unstabilized inlets. In the first two years more than 70 lawsuits protested the Act as an invalid seizure of private property without compensation. Implementation has been difficult, political pressure intense, and results mixed.

Remember, only 28 of 280 barrier islands identified before World War II remain undeveloped, and less than a fourth of original natural wetlands remain undeveloped to some degree. Focus Study 13.1,

"An Environmental Approach to Shoreline Planning," discusses the basics for preparing a "Do" map and a "Don't" map for a portion of the New Jersey shore. Ian McHarg developed the "McHargian" style of planning with meticulous hand-rendered maps of various spatial components.

These components were then synthesized onto a master map identifying regions for protection or development, lowest impact routes for a transmission line, areas for preservation due to unique attributes or environmental characteristic, etc. One of his books is still available in bookstores: *Design with Nature*, Garden City, NY: Doubleday, 1969. Lewis Mumford wrote the introduction to *Design with Nature* in these words, "...I would put it on the same shelf that contains as yet only a handful of works in a similar vein...classics such as those by Henry Thoreau, George Perkins Marsh, Patrick Geddes, Carl Sauer, Benton MacKaye, and Rachel Carson." The root of today's GIS revolution is in McHarg's work as described in this book. For instance examine the thorough geographic analysis and work up of Staten Island (pp. 102–113). Many of you will find a 1970 film by McHarg, produced by WGBH Boston with assistance from Narendra Juneja, called "Multiply and Subdue the Earth" (two reels), in your local educational film library—an old but prophetic film!

Overhead Transparencies

239.	13.1	Principal oceans and seas of the world
240.	13.2	Physical characteristics of the ocean (bottom)
241.	13.3	The littoral zone
242.	13.4	Satellite image of worldwide ocean topography
243.	13.5	Causes of the tides (Sun, Moon, and Earth relationships)
244.	13.7 a, b	Wave formation illustration (top); breakers along the California coast (bottom)
245.	13.8a, b, c, d	Coastal straightening
246.	13.9 a, b	Longshore currents
247.	13.10a, b, c, d	Characteristic erosional coastal features
248.	13.11a, b, c	Characteristic depositional coastal features
249.	13.12 a	Interfering with littoral drift with constructions
250.	13.13 a	Satellite image of a barrier island chain and locator map
251.	13.14 a, b, c	Topo map Chandeleur Islands (top); Chandeleur Island composite photo (middle); *GOES-8* of Hurricane Georges (bottom)
252.	13.15 (top), 13.16a, b (bottom)	Worldwide distribution of corals (map, top); coral forms (art, middle); image of an atoll in the Maldive Islands (bottom)
253.	F.S. 13.1.1a–h	Coastal environment, a planning perspective

14

Glacial and Periglacial Landscapes

A large measure of the freshwater on Earth is frozen, with the bulk of that ice sitting restlessly in just two places—Greenland and Antarctica. The remaining ice covers various mountains and fills alpine valleys. More than 29 million km³ (7 million mi³), or about 77% of all freshwater, is tied up as ice. These deposits of ice, laid down over several million years, provide an extensive frozen record of Earth's climatic history and perhaps some clues to its climatic future. The inference is that rather than distant frozen places of low population, these frozen lands are dynamic and susceptible to change, just as they have been over Earth's past history. The changes in the ice mass worldwide signal vast climatic change and further glacio-eustatic increases in sea level. As you study this chapter, keep this significance in mind.

Elemental Geosystems presents the many recent changes occurring at high latitude and polar locations. For example: collapsing and disintegrating ice shelves around Antarctica, the disappearance of almost half of Arctic Ocean ice since 1970, and increasing depth to the active layer in permafrost regions of Canada and Europe leading to failure of structures and roadbeds built on them.

As an example, please consult Mark F. Meier (U.S. Geological Survey, Tacoma, Washington), "Contribution of Small Glaciers to Global Sea Level," *Science* Vol. 226 (21 December 1984): 1416–1421. He states in his summary that

> Observed long-term changes in glacier volume and hydrometeorological mass balance models yield data on the transfer of water from glaciers, excluding those in Greenland and Antarctica, to the oceans....These glaciers appear to account for a third to half of observed rise in sea level, approximately that fraction not explained by thermal expansion of the ocean.

Outline Headings and Key Terms

The first-, second-, and third-order headings that divide Chapter 14 serve as an outline. The key terms and concepts that appear **boldface** in the text are listed here under their appropriate heading in ***bold italics***. All these highlighted terms appear in the text glossary. Note the check-off box (❏) so you can mark class progress. Your students have this same outline in their Student Study Guide. The ✆ icon indicates that there is an accompanying animation or other resource on the CD.

The outline headings for Chapter 14:

❏ *cryosphere*

Rivers of Ice

❏ *glacier*

Alpine Glaciers

❏ *alpine glacier*

❏ *cirque*

❏ *icebergs*

Continental Glaciers

❏ *continental glacier*

❏ *ice sheet*

❏ *ice cap*

❏ *ice field*

Glacial Processes

✆ **Budget of a Glacier, Mass Balance**

✆ **Flow of Ice Within a Glacier**

Formation of Glacial Ice

❏ *firn*

❏ *glacial ice*

Glacial Mass Balance

❏ *firn line*

❏ *ablation*

❏ *equilibrium line*

Glacial Movement

❏ *crevasses*

Glacial Surges

The URLs related to this chapter of *Elemental Geosystems* can be found at **http://www.prenticehall.com/christopherson**

Key Learning Concepts

After reading the chapter and using this study guide, the student should be able to:

• *Differentiate* between alpine and continental glaciers and *describe* their principal features.

• *Describe* the process of glacial ice formation and *portray* the mechanics of glacial movement.

• *Describe* characteristic erosional and depositional landforms created by alpine glaciation and continental glaciation.

• *Analyze* the spatial distribution of periglacial processes and *describe* several unique landforms and topographic features related to permafrost and frozen ground phenomena.

• *Explain* the Pleistocene ice-age epoch and related glacials and interglacials and *describe* some of the methods used to study paleoclimatology.

Annotated Chapter Review Questions

• *Differentiate* **between alpine and continental glaciers and** *describe* **their principal features.**

1. Describe the location of most freshwater on Earth today. What is the cryosphere?

A large measure of the freshwater on Earth is frozen, with the bulk of that ice sitting restlessly in just two places—Greenland and Antarctica. The remaining ice covers various mountains and fills alpine valleys. More than 29 million km^3 (7 million mi^3) of water, or about 77% of all freshwater, is tied up as ice. Earth's **cryosphere** is the portion of the hydrosphere and ground that is perennially frozen, generally at high latitudes and elevations.

2. What is a glacier? What is implied about existing climate patterns in a glacial region?

A glacier is a large mass of perennial ice, resting on land or floating shelf-like in the sea adjacent to land. Glaciers form by the accumulation and recrystallization of snow. They move under the pressure of their own mass and the pull of gravity. Today, about 11% of Earth's land area is dominated by these slowly flowing ice streams. During colder episodes in the past, as much as 30% of continental land was covered by glacial ice because below-freezing temperatures prevailed at lower latitudes, allowing snow to accumulate. Relative to elevation, in equatorial mountains, the snowline is around 5000 m (16,400 ft); on midlatitude mountains, such as the European Alps, snowlines average 2700 m (8850 ft); and in southern Greenland snowlines are down to 600 m (1970 ft).

3. Differentiate between an alpine glacier and a continental glacier.

With few exceptions, a glacier in a mountain range is called an alpine glacier, or mountain glacier. It occurs in several subtypes. One prominent type is a valley glacier, an ice mass constricted within the confines of a valley. Such glaciers range in length from only 100 m (325 ft) to over 100 km (62 mi). The snowfield that feeds the glacier with new snow is at a higher elevation. As a valley glacier flows slowly downhill, the mountains, canyons, and river valleys beneath its mass are profoundly altered by its passage. A continuous mass of ice is known as a continental glacier and in its most extensive form is called an ice sheet. Two additional types of continuous ice cover associated with mountain locations are designated as ice caps and ice fields. Most glacial ice exists in the snow-covered ice sheets that blanket 80% of Greenland (1.8 million km^3, or 0.43 million mi^3) and 90% of Antarctica (13.9 million km^3, or 3.3 million mi^3).

4. Name the three types of continental glaciers. What is the basis for dividing continental glaciers into types? Which type covers Antarctica?

The three types are ice sheet, ice cap, and ice field. Both ice caps and ice sheets completely bury the underlying landscape, although an ice cap is somewhat circular and covers an area of less than 50,000 km^2 (19,300 mi^2). Antarctica alone has 91% of all the glacial ice on the planet as an enormous ice sheet, or more accurately several ice sheets acting in concert. An ice field is the smallest of the three.

• *Describe* the process of glacial ice formation and *portray* the mechanics of glacial movement.

5. Trace the evolution of glacial ice from fresh fallen snow.

Snow which survives the summer and lasts into the following winter begins a slow transformation into glacial ice. Air spaces among ice crystals are pressed out as snow packs to a greater density. The ice crystals recrystallize under pressure and go through a process of re-growth and enlargement. In a transition step to glacial ice, snow becomes firn, which has a granular texture. As this process continues, many years pass before denser glacial ice is produced. Formation of glacial ice is analogous to formation of metamorphic rock: sediments (snow and firn) are pressured and recrystallized into a dense metamorphic rock (glacial ice).

The formation of glacial ice is analogous to formation of a metamorphic rock with sediments (snow and firn) pressured and recrystallized into a dense metamorphic rock (glacial ice). The time taken for such a process is dependent on climatic conditions as noted in the text. A dry climate such as that found in Antarctica may produce glacial ice in 1000 years, whereas a wetter climate may take only a few years to complete the formation process.

The term névé is a French term for a mass of hardened snow at the head or in the accumulation area of the glacier. When you see this term used interchangeably with firn, that stems from an earlier English application. The American Geological Institute *Glossary of Geology* (3rd edition) recommends that we restrict the usage of névé to refer only to the geographic area covered with perennial snow, or the accumulation area at the head of the glacier. Firn on the other hand is representative of the intermediate stage between snow and glacial ice.

6. What is meant by glacial mass balance? What are the basic inputs and outputs underlying that balance?

A glacier is fed by snowfall and is wasted by losses from its upper and lower surfaces and along its margins. A snowline called a firn line is visible across the surface of a glacier, indicating where the winter snows and ice accumulation survived the summer melting season. A glacier's area of excessive accumulation is, logically, at colder, higher elevations. The zone where accumulation gain ends and loss begins is the equilibrium line. This area of a glacier generally coincides with the firn line, except in subpolar glaciers, where frozen surface water occurs below the firn line. Glaciers achieve a positive net balance of mass—grow larger—during colder periods with adequate precipitation. In a glacier's lower elevation, losses of mass occur because of surface melting, internal and basal melting, sublimation, wind removal by deflation, and the calving, or breaking off, of ice blocks. In warmer times, the equilibrium line migrates to a higher elevation and the glacier retreats—grows smaller—due to its negative net balance.

In data provided to the author by the Ice and Climate Project, University of Puget Sound, Tacoma, Washington 98416, by Robert M. Krimmel, U.S. Geological Survey, it is interesting to note the net mass balance for the South Cascade Glacier, Washington, for the period 1955 through 1998. To update this analysis a new report is now available: Robert M. Krimmel, *Water, Ice, Meteorological, and Speed Measurements at South Cascade Glacier, Washington, 2002 Balance Year*, USGS Water Resources Report, Tacoma, WA: USGS, 2002.

You may want to have your students plot these positive and negative values on a graph, and determine the algebraic sum of the net balance values. The glacier has gone through a net wastage at lower and middle elevations. (Note that this is presented in the Student Study Guide and in the *Applied Physical Geography* lab manual.) These statistics are graphically illustrated in News Report 14.1, documenting the net mass balance for the South Cascade Glacier in Washington state between 1955 and 1998.

The glacier has gone through a net wastage between 1955 and 1998 at lower and middle elevations. In just one year (9/1991 to 10/1992), the terminus retreated 38 m and resulted in other major changes in the glacier. This represents more than a 2% loss in the glacier's mass in that time frame. Net mass balance is specified as cm of water equivalent spread over the entire glacier.

South Cascade Glacier
Annual Net Mass
Balance Data (cm)

Year	Net balance
1955	+30.00
1956	+20.00
1957	−20.00
1958	−330.00
1959	+70.00
1960	−50.00
1961	−110.00
1962	+20.00
1963	−130.00
1964	+120.00
1965	−17.00
1966	−103.00
1967	−63.00
1968	+1.00
1969	−73.00
1970	−120.00

1971	+60.00
1972	+143.00
1973	−104.00
1974	+102.00
1975	−5.00
1976	+95.00
1977	−130.00
1978	−38.00
1979	−156.00
1980	−102.00
1981	−84.00
1982	+8.00
1983	−77.00
1984	+12.00
1985	−120.00
1986	−71.00
1987	−206.00
1988	−164.00
1989	−71.00
1990	−73.00
1991	−20.00
1992	−201.00
1993	−123.00
1994	−160.00
1995	−69.00
1996	+10.00
1997	+63.00
1998	−186.00
1999	+102.00
2000	+38.00
2001	-157.00
2002	+55.00
2003	-165.00
2004	-210.00

7. What is meant by a glacial surge? What do scientists think produces surging episodes?

Some glaciers will lurch forward with little or no warning in a glacial surge. This is not quite as abrupt as it sounds; in glacial terms, a surge can be tens of meters per day. The Jakobshavn Glacier in Greenland, for example, is known to move between 7 and 12 km (4.3 and 7.5 mi) a year. The exact cause of such a glacial surge is still being studied. Some surge events result from a buildup of water pressure in the basal layers of the glacier. Sometimes that pressure is enough to actually float the glacier slightly during the surge. As a surge begins, icequakes are detectable, and ice faults are visible along the margins that separate the glacier from the surrounding stationary terrain.

———————

• *Describe* **characteristic erosional and depositional landforms created by alpine glaciation and continental glaciation.**

8. How does a glacier accomplish erosion?

Glacial erosion is similar to a large excavation project, with the glacier hauling debris from one site to another for deposition. As rock fails along joint planes, the passing glacier mechanically plucks the material and carries it away. There is evidence that rock pieces actually freeze to the basal layers of the glacier and, once embedded, allow the glacier to scour and sandpaper the landscape as it moves, a process called abrasion. This abrasion and gouging produces a smooth surface on exposed rock, which shines with glacial polish when the glacier retreats. Larger rocks in the glacier act much like chisels, working the underlying surface to produce glacial striations parallel to the flow direction.

9. Describe the evolution of a V-shaped stream valley to a U-shaped glaciated valley. What features are visible after a glacier retreats?

W. M. Davis characterized the stages of a valley glacier in a set of drawings published in 1906 and redrawn here in Figure 14.8. Illustration (a) shows a typical river valley with characteristic V-shape and stream-cut tributary valleys that exist before glaciation. Illustration (b) shows that same landscape during a later period of active glaciation. Glacial erosion and transport are actively removing much of the regolith (weathered bedrock) and the soils that covered the preexisting valley landscape. Illustration (c) shows the same landscape at a later time when climates have warmed and ice has retreated. The glaciated valleys now are U-shaped, greatly changed from their previous stream-cut form. You can see the oversteepened sides, the straightened course of the valley, and the presence of hanging valleys and waterfalls. The physical weathering associated with a freeze-thaw cycle has loosened much rock along the steep cliffs, falling to form talus cones along the valley sides during the postglacial period. See Figure 14.8c for labeled details of the features that are formed as a result of the formation, growth, passage, and retreat of an alpine glacier.

10. How is an iceberg generated?

Where a glacier ends in the sea, large pieces break off and drift away as icebergs. These are portions of a glacier at drift in the sea. When large pieces of an ice shelf break off, such as the portions of the Ross ice shelf in the west Antarctic area have done during the past few years, enormous tabular islands are formed as a type of iceberg.

11. Differentiate between two forms of glacial drift—till and outwash.

Where the glacier melts, debris accumulates to mark the former margins of the glacier—the end and sides. Glacial drift is the general term for all glacial deposits. Direct deposits appear unstratified and unsorted and are called till. In contrast, sorted and stratified glacial drift, characteristic of stream-deposited material, is called outwash and forms an outwash plain of glacio-fluvial deposits across the landscape.

12. What is a morainal deposit? What specific moraines are created by alpine and continental glaciers?

Glacial till moving downstream in a glacier can form a marginal unsorted deposit known as a moraine—Figure 14.5 and 14.11. A lateral moraine forms along each side of a glacier. If two glaciers with lateral moraines join, their point of contact becomes a medial moraine. Eroded debris that is dropped at the glacier's farthest extent is called a terminal moraine. However, there also may be end moraines, formed wherever a glacier pauses after reaching a new equilibrium. If a glacier is in retreat, individual deposits are called recessional moraines. And finally, a deposition of till generally spread across a surface is called a ground moraine.

13. What are some common depositional features encountered in a till plain?

With the retreat of the glaciers, many relatively flat plains of unsorted coarse till were formed behind terminal moraines. Low, rolling relief and deranged drainage patterns are characteristic of these till plains. As the glacier melts, this unsorted cargo of ablation till is lowered to the ground surface, sometimes covering the clay-rich lodgement till deposited along the base. The rock material is poorly sorted and is difficult to cultivate for farming, but the clays and finer particles can provide a basis for soil development.

14. Contrast a roche moutonnée and a drumlin regarding appearance, orientation, and the way each forms.

Two landforms created by glacial action are streamlined hills, one erosional (called a roche moutonnée) and the other depositional (called a drumlin). A roche moutonnée is an asymmetrical hill of exposed bedrock. Its gently sloping upstream side (stoss side) has been polished smooth by glacial action, whereas its downstream side (lee side) is abrupt and steep where rock was plucked by the glacier (Figure 14.14). A drumlin is deposited till that has been streamlined in the direction of continental ice movement, blunt end upstream and tapered end downstream. Multiple drumlins (called swarms) occur in fields in New York and Wisconsin, among other areas. Sometimes their shape is that of an elongated teaspoon bowl lying face down (see Figure 14.15, photo and topo map).

• *Analyze* the spatial distribution of periglacial processes and *describe* several unique landforms and topographic features related to permafrost and frozen ground phenomena.

15. In terms of climatic types, describe the areas on Earth where periglacial landscapes occur. Include both higher latitude and higher altitude climate types.

Periglacial regions occupy over 20% of Earth's land surface (Figure 14.16). The areas are either near permanent ice or are at high elevation, and have ground that is seasonally snow free. Under these conditions, a unique set of periglacial processes operate, including permafrost, frost action, and ground ice.

Climatologically, these regions are in *subarctic* and *polar* climates (especially *tundra*). Such climates occur either at high latitude (tundra and boreal forest environments) or high elevation in lower-latitude mountains (alpine environments).

16. Define two types of permafrost, and differentiate their occurrence on Earth. What are the characteristics of each?

When soil or rock temperatures remain below 0°C (32°F) for at least two years, a condition of permafrost develops. An area that has permafrost but is not covered by glaciers is considered periglacial. Note that this criterion is based solely on temperature and not on whether water is present. Other than high latitude and low temperatures, two other factors contribute to permafrost: the presence of fossil permafrost from previous ice-age conditions and the insulating effect of snow cover or vegetation that inhibits heat loss.

Permafrost regions are divided into two general categories, continuous and discontinuous, that merge along a general transition zone. Continuous permafrost describes the region of the most severe cold and is perennial, roughly poleward of the −7°C (19°F) mean annual temperature isotherm. Continuous permafrost affects all surfaces except those beneath deep lakes or rivers in the areas shown in Figure 14.15. Continuous permafrost may exceed 1000 m in depth (over 3000 ft) averaging approximately 400 m (1300 ft).

17. Describe the active zone in permafrost regions, and relate the degree of development to specific latitudes.

The active layer is the zone of seasonally frozen ground that exists between the subsurface permafrost layer and the ground surface. The active layer is subjected to consistent daily and seasonal freeze-thaw cycles. This cyclic melting of the active layer affects as little as 10 cm depth in the north (Ellesmere Island, 78° N), up to 2 meters in the southern margins (55° N) of the periglacial region, and 15 m in the alpine permafrost of the Colorado Rockies (40° N). See Figure 14.17.

18. What is a talik? Where might you expect to find taliks and to what depth do they occur?

A talik (derived from a Russian word) is an unfrozen portion of the ground that may occur above, below, or within a body of discontinuous permafrost or beneath a body of water in the continuous region. Taliks are found beneath deep lakes and may extend to bedrock and noncryotic soil under large deep lakes (Figure 14.17). Taliks form connections between the active layer and groundwater, whereas in continuous permafrost groundwater is essentially cut off from water at the surface. In this way, permafrost disrupts aquifers and causes water supply problems.

19. What is the difference between permafrost and ground ice?

In regions of permafrost, subsurface water that is frozen is termed ground ice. The moisture content of areas with ground ice may very from nearly absent in regions of drier permafrost to almost 100% in saturated soils. From the area of maximum energy loss, freezing progresses through the ground along a freezing front, or boundary between frozen and unfrozen soil. The presence of frozen water in the soil initiates geomorphic processes associated with frost action and the expansion of water volume as it freezes (Chapters 5 and 10).

20. Describe the role of frost action in the formation of various landforms in the periglacial region.

The 9% expansion of water as it freezes produces strong mechanical forces that fracture rock and disrupt soil at and below the surface. Frost-action shatters rock, producing angular pieces that form a block field, or felsenmeer, accumulating as part of the arctic and alpine periglacial landscape, particularly on mountain summits and slopes.

If sufficient water undergoes the phase change to ice, the soil and rocks embedded in the water are subjected to frost-heaving (vertical movement) and frost-thrusting (horizontal motions). Boulders and slabs of rock generally are thrust to the surface. Soil horizons may appear disrupted as if stirred or churned by frost action, a process termed cryoturbation. Frost action also produces a contraction in soil and rock, opening up cracks for ice wedges to form. Also, there is a tremendous increase in pressure in the soil as ice expands, particularly if there are multiple freezing fronts trapping unfrozen soil and water between them.

21. Relate some of the specific problems humans encounter in developing periglacial landscapes.

Human populations in areas that experience frozen ground phenomena encounter various difficulties. Because thawed ground in the active layer above the permafrost zone frequently shifts in periglacial environments, the maintenance of roadbeds and railroad tracks is a particular problem. In addition, any building placed directly on frozen ground will begin to melt itself into the defrosting soil. Thus, the melting of permafrost can create subsidence in structures and complete failure of building integrity (Figure 14.21).

• *Explain* the Pleistocene ice age epoch, and related glacials and interglacials; and *describe* some of the methods used to study paleoclimatology.

22. What is paleoclimatology? Describe Earth's past climatic patterns. Are we experiencing a normal climate pattern in this era or have scientists noticed any significant trends?

Paleoclimatology is the science of past climates. The most recent episode of cold climatic conditions began about 1.65 million years ago, launching the Pleistocene epoch. At the height of the Pleistocene, ice sheets and glaciers covered 30% of Earth's land area, amounting to more than 45 million km^2 (17.4 million mi^2). The Pleistocene is thought to have been one of the more prolonged cold periods in Earth's history. At least 18 expansions of ice occurred over Europe and North America, each obliterating and confusing the evidence from the one before.

The term ice age is applied to any such extended period of cold, even though an ice age is not a single cold spell. Instead, it is a period of generally cold climate, called a glacial, interrupted by brief warm spells, known as interglacials. There is a worldwide retreat of alpine glaciers, higher snowlines in Greenland, and at least three accelerating ice streams on the West Antarctic ice sheet. The reduction in alpine glacial mass balances is particularly true of low and middle elevation glaciers. This trend in ice mass reduction may be attributed to the present century-long increase in mean global air temperatures. Additionally, over the past ten years, we have experienced the warmest years in instrumental history.

23. Define an ice age. When was the most recent? Explain "glacial" and "interglacial" in your answer.

The term ice age is applied to any extended period of cold, even though an ice age is not a single cold spell. Instead, it is a period of generally cold climate, called a glacial, interrupted by brief warm spells, known as interglacials. Traditionally, four major glacials and three interglacials were acknowledged for the Pleistocene epoch. (The glacials were named the Nebraskan, Kansan, Illinoian, and Wisconsinan.) In Europe, similar episodes coincided with those in North America, but were given different names. Modern techniques have opened the way for a new chronology and understanding. Currently, glaciologists acknowledge the Illinoian glacial and Wisconsinan glacial, and the Sangamon interglacial between them. These span the past 300,000 years (Figure 14.23). The Illinoian is believed to have had two glacials (designated stages 6 and 8 in the figure), as did part of the Wisconsinan (stages 2 and 4), which is dated at 10,000 to 35,000 years ago. The stages on the chart are numbered back to stage 23, at approximately 900,000 years ago. The Holocene (past 10,000 years) is regarded as either an interglacial or a post-glacial epoch.

24. Summarize what science has learned about the causes of ice ages by listing and explaining at least four possible factors in climate change.

The mechanisms that bring on an ice age are the subject of much research and debate. Because past occurrences of low temperatures appear to have followed a pattern, researchers have looked for causes that also are cyclic in nature. They have identified a complicated mix of interacting variables that appear to influence long-term climatic trends, including galactic and Earth-Sun relationships, solar variability, geophysical factors, and geographical-geological factors.

25. Describe the role of ice cores in deciphering past climates. What record do they preserve? Where were they drilled?

Ice core analysis has opened the way for a new chronology and understanding of glaciation. Chemical and physical properties of the atmosphere and snow that accumulated each year are frozen into place. Locked into the ice cores are the air bubbles from past atmospheres, which indicate ancient gas concentrations, such as greenhouse gases, carbon dioxide and methane. For example, during cold periods, high concentrations of dust are present, brought by winds from distant dry lands, acting as condensation nuclei. Past volcanic eruptions are recorded in this manner. The presence of ammonia indicates ancient forest fires at lower latitudes and the ratio between stable forms of oxygen is measured with each snowfall.

Greenland has been the location of two ice cores, the Greenland Ice Core Project (GRIP) which began in 1989, and more recently, the Greenland Ice Sheet Project (GISP-2) began in 1990. The GRIP catalogued 250,000 years of history, drilling a 3030m deep core, and the GISP-2 core reached bedrock at a

2700 m and records 115,000 years of past climatic history (see Figure 14.1.1, Focus Study 14.1).

On the East Antarctic Plateau, 75.1° S, 123.3° E, elevation 3233 m (10,607 ft), approximately 900 km (560 mi) from the coast and 1750 km (1087 mi) from the South Pole, sits Dome C. This is the point on the Antarctic continent where the ice sheet is the thickest (see Figure 14.29c for location; and Figure 14.1.2). Here the ice is 3309 m ± 22 m deep (10,856 ft). The mean annual surface temperature is –54.5°C (–66.1°F); however, the science crews experience temperatures that range from –50 C° (–58 F°) when they arrive to –25 C° (–13 F°) by midsummer. Dome C is 560 km (348 mi) from the Vostok base, where the previous coring record of 400,000 years was set. The Dome-C 8-year project is part of a 10-country European Project for Ice Coring in Antarctica (EPICA).

The ice core provides a high-resolution record of ancient atmospheric gases, ash from volcanic eruptions, and materials from other atmospheric events. Through 2003, coring to a 3190-m (2-mi) depth has brought up 740,000 years of Earth's past climate history (actually, 807,000 years has been retrieved but not yet analyzed) (Figure 14.1.3). Scientists analyzing the results, report that the Dome-C record affirms the Vostok findings and correlates perfectly with the deep-sea core of oxygen isotope fluctuations in foraminifera shells (microfossils) from the Atlantic Ocean. Confirmed is the finding that the present concentration of carbon dioxide in the atmosphere is the highest it has been in the past 440,000 years and that these changing levels have marched in step with higher and lower temperatures throughout this time span.

26. Explain the relationship between the criteria defining the Arctic and Antarctic regions. Is there any coincidence in these criteria and the distribution of Northern Hemisphere forests on the continents?

Climatologists use environmental criteria to define the Arctic and the Antarctic regions (Figure 14.26). For the Arctic area, the 10°C (50°F) isotherm for July is used; it coincides with the visible treeline, which is the boundary between the northern forests and tundra climates, (i.e., a temperature below which boreal forests cannot survive).

The *Antarctic region* is defined by the *antarctic convergence*, a narrow zone that extends around the continent as a boundary between colder antarctic water and warmer water at lower latitudes. This boundary follows roughly the 10°C (50°F) isotherm for February (Southern Hemisphere summer) and is located near 60° S latitude. The Antarctic region that is covered with sea ice represents an area greater

than North America, Greenland, and Western Europe combined!

27. What is the latest news fro the analysis of the Dome C ice core as described in the text?.

The term ice age is applied to any extended period of cold, even though an ice age is not a single cold spell. Instead, it is a period of generally cold climate, called a glacial, int

28. What is the significance of having the Dome C ice core include Marine Isotope Stage 11 (MIS 11) shown on Figure 14.23, relative to understanding present climates?

This warm spell at MIS 11, from 425,000 to 395,000 years ago, is perfectly recorded in the Dome C core and is of interest because CO2 in the atmosphere during MIS 11 was similar to our pre-industrial level of 280 ppm. Earth's orbital alignment was similar to that of today as well. The MIS 11 analogy means that a return to the next glacial interval is perhaps 16,000 years in the future.

29. Briefly characterize changes in the cryosphere, specifically in the Arctic and Antarctica regions, that scientists are tracking.

Review the "Global Climate Change" section in Chapter 7, where a portrait of high-latitude temperatures and ongoing physical changes to pack ice, ice shelves, and glaciers is presented. News Report 14.3 looks at meltpond occurrence as indicators of changing surface energy budgets. Throughout this text, several High Latitude Connection features described additional dynamics in these polar regions. High Latitude Connection 14.1 covers the Ward Hunt Ice Shelf breakup. See also HLCs 1.1, 3.1, 8.1, and 17.1.

30. What happened to the Ward Hunt Ice Shelf in northern Canada?

By late fall 2003 the Ward Hunt Ice Shelf—largest in the Arctic, on the north coast of Ellesmere Island in Nunavut—was breaking up. This shelf had been stable for at least 4500 years. After three decades of mass losses, the ice shelf reached a systems threshold and rapidly broke up between 2000 and 2003 (Figure 14.1.1).

Overhead Transparencies

254.	14.1	Alpine glaciers, Ellesmere Island satellite detail
255.	14.2	Ruth and Eldridge glaciers in south-central Alaska
256.	14.3	Image of Eugenie Glacier icebergs break off in Dobbins Bay
257.	14.5 a, b, d	Retreating alpine glacier (a) and mass balance (b), (d) satellite image of rivers of ice detail (bottom)
258.	14.6 a, b, c	Glacier cross section showing movement dynamics; surface crevasses (middle); Pine Island Glacier at bottom.
259.	14.8 a, b, c	Geomorphic handiwork of alpine glaciers (before, during, and after a valley glacier occupies a formerly stream-cut valley)
260.	14.11 a, b, c	Continental glacier depositional features
261.	14.12a, b	Roche moutonnée, Lembert Dome; development of a roche moutonnée
262.	14.14	Distribution of permafrost in Northern Hemisphere
263.	14.15	Cross section of typical periglacial region in Northern Canada
264.	14.18	Building failure/melting permafrost
265.	14.21a, b	Pleistocene glaciation (polar projection with blow-out map of North America 18,000 years ago and smaller map form 9500 years ago)
266.	14.22,a, b, c, d	Paleolakes of the western U.S.
267.	14.23 a, b, and c	Recent climates from ice cores (top); GISP 2 record 500 years to the present (bottom)
268.	14.24a, b, c	Astronomical factors that affect climate cycles
269.	14.26a, b	The Arctic and Antarctic regions
270.	14.35	Amundsen-Scott Base, South Pole
271.	N.R. 14.1.1	South Cascade Mass Balance graph
272.	N.R. 14.1.2a, b	Alaskan glaciers mass balances

PART FOUR:
Biogeography

Overview

Earth is the home of the only known biosphere in the Solar System—a unique, complex, and interactive system of abiotic and biotic components working together to sustain a tremendous diversity of life. Thus we begin Part Four of *Elemental Geosystems* and an examination of the biogeography of the biosphere–soils, ecosystems, and terrestrial biomes. Remember the description of the biosphere from Chapter 1:

> The intricate, interconnected web that links all organisms with their physical environment is the **biosphere**. Sometimes called the **ecosphere**, the biosphere is the area in which physical and chemical factors form the context of life. The biosphere exists in the overlap among the abiotic spheres, extending from the seafloor to about 8 km (5 mi) into the atmosphere. Life is sustainable within these natural limits. In turn, life processes have powerfully shaped the other three spheres through various interactive processes. The biosphere has evolved, reorganized itself at times, faced extinction, gained new vitality, and managed to flourish overall. Earth's biosphere is the only one known in the Solar System; thus, life as we know it is unique to Earth.

Part Four is a synthesis of many of the elements covered throughout the text.

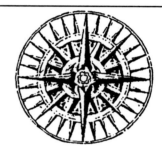

The Geography of Soils

Earth's landscape is generally covered with soil. Soil is a dynamic natural body comprises fine materials in which plants grow, and which is composed of both mineral and organic matter. By their diverse nature, soils are a complex subject and pose a challenge for spatial analysis. This chapter presents an overview of the modern system of soil classification used in the United States, with mention of the Canadian system. The Soil Taxonomy system is often misnamed in texts, and anachronistic soil classification terminology is sometimes used. Since the system has been in constant use since 1975 and is integrated into all major pedology textbooks, I present this system in an accurate version true to original source material. The latest revision to the Soil Taxonomy occurred in 1998–99 and *Elemental Geosystems* presents this, including the newest soil order (as of November 1998), the Gelisols.

Please find Table 15.1, and utilize it to simplify this geographic study of soils. The table summarizes the 12 soil orders of the Soil Taxonomy, including general location and climate association, areal coverage estimate, and basic description. Also, you will find small locator maps for most soil orders included with a picture of that soil's profile. These locator maps are derived from the new worldwide distribution map presented in Figure 15.8. Please consult the companion text *Geosystems—An Introduction to Physical Geography*, sixth edition 2006, for more detail on soil characteristics including "Principal Soil Moisture Regimes," Table 18.2. A description and analysis of the Canadian System of Soil Classification (CSSC), Appendix B, includes a soils map and descriptive tables.

Outline Headings and Key Terms

The first-, second-, and third-order headings that divide Chapter 15 serve as an outline for your notes and studies. The key terms and concepts that appear **boldface** in the text are listed here under their appropriate heading in ***bold italics***. All these highlighted terms appear in the text glossary. Note the check-off box (❑) so you can mark your progress as you master each concept. Your students have this same outline in their Student Study Guide. The ✪ icon indicates that there is an accompanying animation or other resource on the CD.

The outline headings for Chapter 15:
- ❑ *soil*
- ❑ *soil science*

Soil Characteristics

Soil Profiles
- ❑ *pedon*
- ❑ *polypedon*

Soil Horizons
- ❑ *soil horizon*
- ❑ *humus*
- ❑ *eluviation*
- ❑ *illuviation*
- ❑ *solum*

Soil Properties

✪ **Soil Ion Exchange: Soil Particles and Soil Water**

Soil Color

Soil Texture
- ❑ *loam*

Soil Structure

Soil Consistence

Soil Porosity

Soil Moisture

Soil Chemistry

Soil Colloids and Mineral Ions
- ❑ *soil colloids*
- ❑ *adsorption*
- ❑ *cation-exchange capacity (CEC)*
- ❑ *soil fertility*

Soil Acidity and Alkalinity

Soil Formation Factors and Management

Natural Factors

The Human Factor

Soil Classification

Soil Taxonomy
- ❑ ***Soil Taxonomy***

Pedogenic Regimes
- ❑ *laterization*
- ❑ *salinization*
- ❑ *calcification*
- ❑ *podzolization*
- ❑ *gleization*

Diagnostic Soil Horizons
- ❑ *epipedon*
- ❑ *diagnostic subsurface horizon*

The 12 Soil Orders of the Soil Taxonomy

Oxisols
- ❑ *Oxisols*
- ❑ *plinthite*

Aridisols
- ❑ *Aridisols*

Mollisols
- ❑ *Mollisols*

Alfisols
- ❑ *Alfisols*

Ultisols
- ❑ *Ultisols*

Spodosols
- ❑ *Spodosols*

Entisols
- ❑ *Entisols*

Inceptisols
- ❑ *Inceptisols*

Gelisols
- ❑ *Gelisols*

Andisols
- ❑ *Andisols*

Vertisols
- ❑ *Vertisols*

Histosols
- ❑ *Histosols*

Summary and Review
News Report and Focus Study

News Report 15.1: Soil Is Slipping through Our Fingers
Focus Study 15.1: Selenium Concentration in Western Soils

The URLs related to this chapter of *Elemental Geosystems* can be found at
http://www.prenticehall.com/christopherson

Key Learning Concepts

After reading the chapter and using the Student Study Guide, the student should be able to:

- *Define* soil and soil science and *describe* a pedon, polypedon, and typical soil profile.
- *Describe* soil properties of color, texture, structure, consistence, porosity, and soil moisture.
- *Explain* basic soil chemistry, including cation-exchange capacity, and *relate* these concepts to soil fertility.
- *Evaluate* principal soil formation factors, including the human element.
- *Describe* the 12 soil orders of the Soil Taxonomy classification system and *explain* their general occurrence.

Annotated Chapter Review Questions

• *Define* soil and soil science and *describe* a pedon, polypedon, and typical soil profile.

1. Soils provide the foundation for animal and plant life and therefore are critical to Earth's ecosystems. Why is this true?

Soil is a dynamic natural body comprised of fine materials in which plants grow, and which is composed of both mineral and organic matter. Specific soil conditions determine soil fertility, which is the ability of soil to support plant productivity. Plants capture sunlight and fix carbon in organic compounds that sustain the biosphere.

2. What are the differences among soil science, pedology, and edaphology?

Soil science is interdisciplinary, involving physics, chemistry, biology, mineralogy, hydrology, taxonomy, climatology, and cartography. Pedology concerns the origin, classification, distribution, and description of soil. Pedology is at the center of learning about soils, yet is does not dwell on its practical uses. Edaphology focuses on soil as a medium for sustaining higher plants. Edaphology emphasizes plant growth, fertility, and the differences in productivity among soils. Pedology gives us a general understanding of soils and their classification, whereas edaphology reflects society's concern for food and fiber production and the management of soils to increase fertility and reduce soil losses.

3. Define polypedon and pedon, the basic units of soil.

A soil profile selected for study should extend from the surface to the lowest extent of plant roots, or to the point where regolith or bedrock is encountered. Such a profile, known as a pedon, is imagined as a hexagonal column encompassing from 1 m^2 to 10 m^2 in surface area (Figure 15.1). At the sides of the pedon, the various layers of the soil profile are visible in cross section. A pedon is the basic sampling unit in soil surveys. Many pedons together in one area comprise a polypedon, which has distinctive characteristics differentiating it from surrounding polypedons. These polypedons are the essential soil individuals, constituting an identifiable series of soils in an area. A polypedon has a minimum dimension of about 1 m2 and no specified maximum size. It is the soil unit used in preparing local soil maps.

4. Characterize the principal aspects of each soil horizon. Where does the main accumulation of organic material occur? Where does humus form? Explain the difference between the eluviated layer and the illuviated layer. Which horizons constitute the solum?

Each layer exposed in a pedon is a soil horizon. A horizon is roughly parallel to the pedon's surface and has characteristics distinctly different from horizons directly above or below. The boundary between horizons usually is visible in the field, using the properties of color, texture, structure, consistency porosity, the presence or absence of certain minerals, moisture, and chemical processes.

At the top of the soil profile is the O horizon, composed of organic material derived from plant and animal litter that was deposited on the surface and transformed into humus. Humus is a mixture of decomposed organic materials in the soil and is usually dark in color. At the bottom of the soil profile is the R horizon, representing either unconsolidated material or consolidated bedrock of granite, sandstone, limestone, or other rock. The A, B, and C horizons mark differing mineral strata between O and R; these middle layers are composed of sand, silt, clay, and other weathered by-products. In the A horizon, the presence of humus and clay particles is particularly important, for they provide essential chemical links between soil nutrients and plants.

The lower portion of the A horizon grades into the E horizon, which is a bit more pale and is made up of coarse sand, silt, and resistant minerals. Clays and oxides of aluminum and iron are leached (removed) from the E horizon and migrate to lower horizons with water as it percolates through the soil. This process of rinsing through upper horizons and removing finer particles and minerals is termed eluviation; thus the designation E for this horizon. The greater the precipitation in an area, the higher the rate of eluviation that occurs in the E horizon.

Materials are translocated to lower horizons by internal washing in the soil. This process of rinsing through upper horizons and removing finer particles and minerals is termed eluviation—an erosional process. In contrast to A horizons, B horizons demonstrate an accumulation of clays, aluminum, iron, and possibly humus. These horizons are dominated by illuviation—a depositional process. The C horizon is weathered bedrock or weathered parent material, excluding the bedrock itself. This zone is identified as regolith.

The combination of A horizon with its eluviation removals and the B horizon with its illuviation accumulations is designated the solum, considered the true soil of the pedon. A and B horizons are most representative of active soil processes.

_____

• *Describe* soil properties of color, texture, structure, consistence, porosity, and soil moisture.

5. How can soil color be an indication of soil qualities? Give a couple of examples.

Color is important, for it sometimes reflects composition and chemical makeup. Soil scientists describe a soil's color by comparing it with a Munsell Color Chart (from the Munsell Color Company). These charts are in a loose-leaf binder, with 175 different colors arranged by hue (the dominant spectral color, such as red), value (degree of darkness or lightness), and chroma (purity and strength of the color saturation, which increases with decreasing grayness). Color is identified by a name and a Munsell notation, and checked at various depths within a pedon.

6. Define a soil separate. What are the various sizes of particles in soil? What is loam? Why is loam regarded so highly by agriculturalists?

Individual mineral particles are called soil separates; those smaller than 2 mm in diameter (0.08 in.), such as very coarse sand, are considered part of the soil, whereas larger particles are identified as pebbles, gravels, or cobbles. Figure 15.3 shows a diagram of soil textures with sand, silt, and clay concentrations. The figure includes the common designation loam, which is a mixture of sand, silt, and clay in almost equal shares. The water-holding characteristics and ease of cultivation make a sandy loam soil, with clay content below 30%, ideal for farmers.

7. What is a quick, hands-on method for determining soil consistence?

A corollary to texture and structure is soil consistency which is the cohesion in soil and its resistance to mechanical stress and manipulation under varying moisture conditions. Wet soils are variably sticky when held between the thumb and forefinger, ranging from a little adherence to either finger, to sticking to both fingers, to stretching when the fingers are moved apart. Plasticity, the quality of being molded, is roughly measured by rolling a piece of soil between your fingers and thumb to see whether it rolls into a thin strand. Moist soil implies that it is filled to about half of field capacity, and its consistence grades from loose (noncoherent), to friable (easily pulverized), to firm (not crushable between thumb and forefinger).

8. Summarize the role of soil moisture in mature soils.

Soil moisture regimes and their associated climate types shape the biotic and abiotic properties of the soil more than any other factor. Based on Thornthwaite's water-balance principles (Chapter 7, Figures 7.7 and 7.8), the U.S. Soil Conservation Service recognizes five soil moisture regimes.

• *Explain* basic soil chemistry, including cation-exchange capacity, and *relate* these concepts to soil fertility.

9. What are soil colloids? How do they relate to cations and anions in the soil? Explain cation-exchange capacity.

Soil colloids are important for retention of ions in soil. These tiny clay and organic particles carry a negative electrical charge and consequently are attracted to any positively charged ions in the soil. Clay colloids and organic colloids exhibit different levels of chemical activity. Individual clay colloids are thin and platelike, with parallel surfaces that are negatively charged. Colloids can exchange cations between their surfaces and the soil solution, a measured ability called cation-exchange capacity (CEC). A high CEC means that the soil colloids can store or exchange more cations from the soil solution, an indication of good soil fertility (unless there is a complicating factor, such as the soil being too acid). Cations attach to the surfaces of the colloids by adsorption, that is, the metallic cations are adsorbed by the soil colloids. A cation is a positively charged ion and an anion is a negatively charged ion.

10. What is meant by the concept of soil fertility?

Soil fertility is the ability of soil to sustain plants. Soil is fertile when it contains organic substances and clay mineral that absorb certain elements needed by plants.

• *Evaluate* principal soil formation factors, including the human element.

11. Briefly describe the contribution of the following factors and their effect on soil formation: parent material, climate, vegetation, landforms, time, and humans.

The role of parent material in providing weathered minerals to form soils is important in establishing the basic mineral structure and character of the developing soil. At the bottom of the soil profile is the R horizon, representing either unconsolidated material or consolidated bedrock of granite, sandstone, limestone, or other rock. When bedrock weathers to form regolith, it may or may not contribute to overlying soil horizons.

Worldwide, soil types show a close correlation to climate types. The moisture, evaporation, and temperature regimes associated with varying climates determine the chemical reactions, organic activity, and eluviation rates of soils. Not only is the present climate important, but many soils also exhibit the imprint of past climates, sometimes over thousands of years.

The organic content of soil is determined in part by the vegetation growing in that soil, as well as by animal and bacterial activity. The chemical makeup of vegetation contributes to acidity or alkalinity in the soil solution. For example, broadleaf trees tend to increase alkalinity, whereas needleleaf trees tend to produce higher acidity.

Landforms also affect soil formation, mainly through slope and orientation. Slopes that are too steep do not have full soil development, but slopes that are slight may inhibit soil drainage. As for orientation, in the Northern Hemisphere, a southern slope exposure is warmest (slope faces the southern Sun), which affects water balance relationships.

All of these identified factors require time to operate. A few centimeters' thickness of prime farmland soil may require 500 years for maturation. Yet these same soils are being lost at a few centimeters per year to sheetwash and gullying produced by human abuse of the soil resource. The GAO estimates that between 3 and 5 million acres of prime farmland are being lost each year in the United States directly because of poor practices.

12. Explain some of the details that support concern over loss of our most fertile soils. What cost estimates have been placed on soil erosion?

Much effort and many dollars are expended to create fertile soil conditions, yet we live in an era when the future of Earth's most fertile soils is threatened. Soil erosion is increasing worldwide. Some 35% of farmland is losing soil faster than it can form–a loss exceeding 22.75 billion metric tons per year (25 billion tons). Increases in production resulting from artificial fertilizers and new crop designs partially mask this effect, but such compensations for soil loss are nearing an end. Soil depletion and loss are at record levels from Iowa to China, Peru to Ethiopia, and the Middle East to the Americas. The impact on society could be significant. One 1995 study tabulated the market value of lost nutrients and other variables at over $25 billion a year in the United States and hundreds of billion dollars worldwide. The cost to bring soil erosion under control in the United States is estimated at approximately $8.5 billion, or about 30 cents on every dollar of damage and loss.

———————

• *Describe* the 12 soil orders of the Soil Taxonomy classification system and *explain* their general occurrence.

13. Summarize what led soil scientists to develop the new Soil Taxonomy classification system?

Taxonomy is defined in Webster's as "Classification according to natural relationships...the systematic distinguishing, ordering, and naming of type groups within a subject field....a formal system of nomenclature." Additional background is added to this section concerning the history of soil classification efforts in Canada.

In 1883, the Russian soil scientist V. V. Dukuchaev published a monograph that organized soils into rough groups based on observable properties, most of which resulted from climatic and biological soil-forming processes. The Russians contributed greatly to modern soil classification because they were first to consider soil an independent natural body with a definite individual soil genesis that was recognizable in an orderly global pattern. They were first to demonstrate broad interrelationships among the physical environment, vegetation, and soils, at a time when a scientific revolution was beginning to sweep their country in various fields. By contrast, American and European scientists were still considering soil as a geologic product, or simply a mixture of chemical compounds.

The first formal American system is credited to Dr. Curtis F. Marbut, who derived his approach from Dukuchaev. Published in the 1935 *Atlas of American Agriculture* and the *Yearbook of Agriculture–1938*, Marbut's classification of great soil groups recognized the importance of soils as products of dynamic natural processes.

The inadequacies of the 1938 system were obvious to soil scientists at about this time. As the number of identified soil series grew, it became increasingly problematic to associate all the soil series to soil families, groups, and higher categories. The 1938 system simply did not contain detailed criteria for the higher categories of classification.

The work of developing a new classification system began about 1950 with a series of approximations, each tested in-progress by the Soil Survey Staff of the Soil Conservation Service in the United States. International cooperation was important, as evidenced by conferences at Ghent, Leopoldville, Paris, London, and Washington. The 3rd Approximation included many of the European suggestions. By 1955, the 4th Approximation was introduced with efforts to group soil series for possible identification of patterns in categories above the soil-family level. This produced the 5th Approximation at the 1956 6th Congress of the International Society of Soil Science in Paris. By 1957 the 6th Approximation was developed to provide the Soil Survey Staff with a guide for better grouping of subgroups and families. Also, through this stage the nomenclature was developed in consultation with specialists in language, the classics, and with particular assistance from Belgian scientists.

By 1960, the 7th Approximation was completed and distributed at the 7th International Congress of Soil Science held at Madison, Wisconsin. At this time we were approaching some 10,000 soil series in active use in the United States. Many texts still use this 30-year old outdated name to describe the Soil Taxonomy system—a practice not duplicated in any soil science or pedology text! In preparing *Elemental Geosystems* I spoke to Dr. Henry D. Foth of Michigan State (*Fundamentals of Soil Science*, 8th ed., Wiley, 1990). He commented on this use of misnomers for Soil Taxonomy and that we should get in line with the related field of soil science. I have made an effort to do this.

Various supplements and updates were subsequently prepared to improve the system especially with regards to improving tropical soil classification. The effort is clearly international in scope and reflects work from many scientists worldwide. All this led to the publication of Agricultural Handbook No. 436 in 1975, the *Soil Taxonomy—A Basic System of Soil Classification for Making and Interpreting Soil Surveys*. Many aspects of Soil Taxonomy overlap with the world soil map published by the Food and Agriculture Organization. If it is not already on your shelf you might want to obtain a copy of this 750-page volume, along with Agricultural Handbook No. 18 the Soil Survey Manual. Also useful is the Soil Management Support Services Technical Monograph No. 6, the *Keys to Soil Taxonomy* by the Soil Survey Staff (3rd ed., 1987). The latest 1998, 1999 revision is described in the text. Over the years various revisions and clarifications in the system were published in *Keys to the Soil Taxonomy*, now in its 8th edition (1998) which include all the revisions to the 1975 Soil Taxonomy. Major revisions include the addition of two new soil orders: Andisols (volcanic soils) in 1990 and Gelisols (cold and frozen soils) in 1998.

Canadian efforts at soil classification began in 1914 with the partial mapping of soils in Ontario by A.J. Galbraith. Efforts to develop a taxonomic system spread across the country, anchored by academic departments at universities in each province. Regional differences emerged, hampered by a lack of specific soil details. Only 1.7% of Canadian soil, totaling 15 million hectares, was surveyed by 1936.

The Canadian System of Soil Classification (CSSC) in Appendix B provides taxa for all soils presently recognized in Canada and is adapted to Canada's expanses of forest, tundra, prairie, frozen ground, and colder climates. As in the U.S. Soil Taxonomy system, the CSSC classifications are based on observable and measurable properties found in real soils rather than idealized soils that may result from the

interactions of genetic processes. The system is flexible in that its framework can accept new findings and information in step with progressive developments in the soil sciences. The system is arranged in a nested, hierarchical pattern to allow generalization at several levels of detail. Elements of the horizon suffix descriptions are derived from the Soil Taxonomy although adapted to conditions in the Canadian environment.

Appendix B, Table 1, is a summary derived from several Canadian publications. I tried to summarize in an admittedly succinct form a large volume of information. Note that the percentage of land area classified under each soil order and the Soil Taxonomy equivalent is given with each characteristics section.

14. What is the basis of the Soil Taxonomy system? How many orders, suborders, great groups, subgroups, families, and soil series are there?

The original version of the Soil Taxonomy system was published in 1975: *Soil Taxonomy—A Basic System of Soil Classification for Making and Interpreting Soil Surveys*, is generally called simply Soil Taxonomy through to its current revision. Much of the information in this chapter is derived from that keystone publication.

The classification system divides soils into six categories, creating a hierarchical sorting system. Each soil series (the smallest, most detailed category) ideally includes only one polypedon, but may include portions continuous with adjoining polypedons in the field. Soil orders = 12; suborders = 47; great groups = 230; subgroups = 1200; families = 6000; and soil series = 15,000.

15. Define an epipedon and a subsurface diagnostic horizon. Give a simple example of each.

Two diagnostic horizons may be identified in the solum: the epipedon and the subsurface. The epipedon (literally, over the soil) is the diagnostic horizon that forms at the surface and may extend through the A horizon, even including all or part of an illuviated B horizon. The epipedon is visibly darkened by organic matter and sometimes is leached of minerals. There are six recognized epipedons. A diagnostic horizon often reflects a physical property (color, texture, structure, consistence, porosity, moisture), or a dominant soil process in a pedon.

The second type of diagnostic horizon is the subsurface diagnostic horizon. It originates below the surface at varying depths and may include part of the A and/or B horizons. Many subsurface diagnostic horizons have been identified

In the Canadian system soil horizons are named and standardized as diagnostic in the classification process. Several mineral and organic horizons and layers are used in the CSSC. Three mineral horizons are recognized by capital letter designation, followed by lowercase suffixes for further description. Principal soil-mineral horizons and suffixes are presented in Appendix B.

16. Locate each soil order on the world map and on the U.S. map as you give a general description of it.

Utilize Table 15.1, for an integrated overview of the twelve major soil orders and their descriptions, characteristics, areal distribution, former equivalent name and Canadian equivalent, and pronunciation. Figure 15.8 presents the worldwide distribution of all 12 soil orders. Small locator maps are also presented for eight of the orders as follows: Figure 15.9 (Oxisols), Figure 15.12 (Aridisols), Figure 15.14 (Mollisols), Figure 15.18 (Alfisols), Figure 15.19 (Ultisols), Figure 15.20 (Spodosols), Figure 15.22 (Gelisols, the newest order), Figure 15.24 (Vertisols), and Figure 15.25 (Histosols). These smaller maps were prepared from Figure 15.8 so that students will be able to relate the individual soil order depicted back to the worldwide pattern.

17. How was slash-and-burn shifting cultivation, as practiced in the past, a form of crop and soil rotation and conservation of soil properties?

Earlier slash-and-burn shifting cultivation practices were adapted to equatorial and tropical soil conditions and formed a unique style of crop rotation. The scenario went like this: people in the tropics cut down (slashed) and burned the rain forest in small tracts, cultivated the land with stick and hoe, and planted maize (corn), beans, and squash. After several years the soil lost fertility, and the people moved on to the next tract to repeat the process. After many years of movement from tract to tract, the group returned to the first patch to begin the cycle again. This practice protected the limited fertility of the soils somewhat, allowing periods of recovery to follow active production. However, the invasion of foreign plantation interests, development by local governments, vastly increased population pressures, and conversion of vast new tracts to pasturage halted this orderly native pattern of land rotation.

18. Describe the salinization process in arid and semiarid soils. What associated soil horizons develop?

A soil process that occurs in Aridisols and nearby soil orders is salinization. Salinization results from excessive POTET rates in the deserts and semiarid regions of the world. Salts dissolved in soil water are brought to surface horizons and deposited there as surface water evaporates. These deposits appear as subsurface salic horizons, which will damage and kill plants when they occur near the root zone. Obviously, salinization complicates farming in Aridisols. The introduction of irrigation water may either water log poorly drained soils or lead to salinization. Nonetheless, vegetation does grow where soils are better drained and lower in salt content.

19. Which of the soil orders are associated with Earth's most productive agricultural areas?

Mollisols (grassland soils) are some of Earth's most significant agricultural soils. There are seven recognized suborders, not all of which bear the same degree of fertility. The dominant diagnostic horizon is called the mollic epipedon, which is a dark, organic surface layer some 25 cm (10 in.) thick. As the Latin name implies, Mollisols are soft, even when dry, with granular or crumbly peds, loosely arranged when dry. These humus-rich organic soils are high in base cations (calcium, magnesium, and potassium) and have a high CEC.

20. What is the significance to plants of the 51 cm (20 in.) isohyet in the Midwest relative to soils, pH, and lime content?

In North America, the Great Plains straddle the 98th meridian, which is coincident with the 51 cm (20 in.) isohyet of annual precipitation—wetter to the east and drier to the west. The Mollisols here mark the historic division between the short- and tall-grass prairies. The relationship among Mollisols, Aridisols (to the west), and Alfisols (to the east) is shown in Figure 15.15. The illustration also presents some of the important graduated changes that denote these different soil regions, including the level of pH concentration and the depth of available lime.

21. Describe the podzolization process associated with northern coniferous forest soils. What characteristics are associated with the surface horizons? What strategies might enhance these soils?

The Spodosols (northern coniferous forest soils) and their four related suborders occur generally to the north and east of the Alfisols. Spodosols lack humus and clay in the A horizons. An eluviated albic horizon, sandy and leached of clays and irons, lies in the A horizon instead and overlies a spodic horizon of illuviated organic matter and iron and aluminum oxides. The surface horizon receives organic litter from base-poor, acid-rich trees, which contribute to acid accumulations in the soil. The low pH (acid) soil solution effectively removes clays, iron, and aluminum, which are passed to the upper diagnostic horizon. An ashen-gray color is common in these subarctic forest soils and is characteristic of a formation process called podzolization. The low base-cation content of Spodosols requires the addition of nitrogen, phosphate, and potash, and perhaps crop rotation as well, if agriculture is to be attempted. A soil amendment such as limestone can significantly increase crop production in these acidic soils.

Podzolic (Russian, podzol) (25 subgroups) Soils formed in association with the conditions of coniferous forests and sometimes heath. Leaching of overlying horizons occurs in response to moist, cool-to-cold climates. Iron, aluminum, and organic matter from L, F, and H horizons are redeposited in podzolic B horizon.

22. What former Inceptisols now form a new soil order named in 1998? Describe these soils as to location, nature, and formation processes. Why do you think they were separated into their own order?

Andisols formerly were considered under Inceptisols and Entisols, but in 1990 they were placed in this new order. Andisols are derived from volcanic ash and glass. Previous soil horizons frequently are found buried by ejecta from repeated eruptions. Volcanic soils are unique in their mineral content and in their recharge by eruptions.

Weathering and mineral transformations are important in this soil order. Volcanic glass weathers readily into allophane (a noncrystalline aluminum silicate clay mineral that acts as a colloid) and oxides of aluminum and iron. Andisols feature a high CEC, high water-holding ability, and develop moderate fertility, although phosphorus availability is an occasional problem.

In November 1998 a new soil order, formerly classified under Inceptisols, Entisols, and Histosols, the Gelisols were designated (Figure 15.22). These are cold and frozen soils associated with high latitude and high elevation conditions.

23. Why has a selenium contamination problem arisen in western soils? Explain the impact of agricultural practices, and tell why you think this is or is not a serious problem.

See Focus Study 15.1. About 95% of the irrigated acreage in the United States is west of the 98th meridian. This region is increasingly troubled by salinization and water-logging problems. But at least nine sites in the West, particularly in California's western San Joaquin valley, are experiencing related contamination of a more serious nature—increasing selenium concentrations. The soils in these areas were derived from former marine sediments that formed shales in the adjoining Coast Ranges. As parent materials weathered, selenium-rich alluvium washed into the semiarid valley, forming the soils that needed only irrigation water to become productive. After 1960, large-scale irrigation efforts intensified, resulting in subtle initial increases in selenium concentrations in soil and water. In trace amounts, selenium is a dietary requirement for animals and humans, but in higher amounts it is toxic to both. Toxic effects were reported during the 1980s in some domestic animals grazing on grasses grown in selenium-rich soils in the Great Plains.

According to U.S. Fish and Wildlife Service scientists, the toxicity moves through the food chain and genetically damages and kills wildlife. For example, birth defects and death were widely reported in all varieties of birds that nested at selenium-contaminated Kesterson Wildlife Refuge; approximately 90% of the exposed birds perished or were injured. Such damage to wildlife presents a real warning to human populations at the top of the food chain. At the very least we must acknowledge that soil processes are complex. Certainly, much remains to be learned to avoid environmental tragedies such as the one at Kesterson—remember that there are nine such threatened sites in the West, and Kesterson was only the first catastrophe.

24. Examine the peat cutting and drying shown in the chapter-opening photo and Figure 15.25c. Explain the soil-forming processes at work to produce these organic soils. How are the peat blocks used? Into which Soil Taxonomy soil order do these soils fall?

Peat , a Histosol organic soil, is cut into blocks by hand with a spade, which then are set out to dry (see this chapter's opening photo). As layer after layer of fibrous material compresses and chemically breaks down, peat is formed. Poor drainage and high-water content is important to the process. In Figure 15.25c, note the fibrous texture of the sphagnum moss growing on the surface and the darkening layers with depth in the soil profile as the peat is compressed and chemically altered. Such beds can be more than 2-m thick. Once dried, the peat blocks burn hot and smoky. Peat is the first stage in the formation of lignite, an intermediate step toward coal. Imagine such soils forming in plant-lush swamp environments in the Carboniferous Period (359 to 299 m.y.a.), only to go through coalification to become coal deposits.

Overhead Transparencies

16

Ecosystems and Biomes

The interaction of the atmosphere, hydrosphere, and lithosphere produces conditions within which the biosphere exists. Chapter 16 begins the process of synthesizing all these "spheres" from throughout *Elemental Geosystems* into a complete spatial picture of Earth that culminates in this chapter. In this complex age, the spatial tools of the geographic approach are uniquely suited to unravel the web of human impact on Earth's systems. Many potential career opportunities are based on a degree in this field, which can lead to planning, GIS applications, environmental impact assessment, and location analysis vocations. The same type of geographic synthesis at a site or environmental workup is the approach of landscape architecture.

The biosphere extends from the floor of the ocean to a height of about 8 km (5 mi) in to the atmosphere. The biosphere is composed of myriad ecosystems from simple to complex, each operating within general spatial boundaries. Ecology is the study of the relationships between organisms and their environment and among the various ecosystems in the biosphere. Biogeography, essentially a spatial ecology, is the study of the distribution of plants and animals and the diverse spatial patterns they create across Earth.

We are the species with developed technology that allows us to lift off the surface of our home planet despite the protests of gravity, to view Earth from afar. At a moment when these accomplishments fill us with pride, we also find ourselves overwhelmed with the immensity of the home planet and the smallness of our everyday reality, as revealed to us by photographs of Earth—such as the one on the back cover of the this textbook, and the image on the front cover. A proper understanding of this global ecosystems chapter is best fueled by an inner (mental) picture of our Earth stimulated by the one taken from space, for it is only from that perspective that the importance of even the smallest community, or the most finite ecosystem, will arise in proper significance. And from such awareness

even the loss of a single species will be a headline and a topic of concern to us all.

To facilitate this process of culmination of various elements from throughout this text, I prepared Table 16.2 to portray aspects of the atmosphere, hydrosphere, and lithosphere that merge to produce the major terrestrial ecosystems. The table is presented in columns covering vegetation characteristics, soil classes, Köppen climate designation, annual precipitation range, temperature patterns, and water balance characteristics.

Focus Study 16.1, "Biodiversity and Biosphere Reserves," presents the biosphere reserve effort to establish protected "islands" of natural biomes. The principal goal is to hold off the record number of extinctions now taking place. Even the most detached individual must recognize that humankind has been an agency of change on Earth—the creators of cultural, albeit artificial, landscapes. The spatial implications of these dynamic trends are of particular importance to physical geographers, for we have the potential for spatial analysis and synthesis. Our discipline is at the heart of geographic information system model construction. The biosphere is quite resilient and adaptable whereas many specific biomes and communities are greatly threatened by further destructive impacts. The irony is that some plants and animals in these biomes contain cures and clues to human disease, potential new food sources, and mechanisms to recycle the excessive levels of carbon dioxide now entering the atmosphere.

Outline Headings and Key Terms

The first-, second-, and third-order headings that divide Chapter 16 serve as an outline for your notes and studies. The key terms and concepts that appear **boldface** in the text are listed here under their appropriate heading in ***bold italics***. All these highlighted terms appear in the text glossary. Note the check-off box (❑) so you can mark your progress as you master each concept. Your students have this same outline in their Student Study Guide. The ⊛ icon indicates that there is an accompanying animation or other resource on the CD.

The outline headings for Chapter 16:
- ❑ *ecosystem*
- ❑ *ecology*
- ❑ *biogeography*

Ecosystem Components and Cycles
⊛ **The Global Carbon Cycle**
⊛ **The Nitrogen Cycle**
⊛ **Midlatitude Productivity**
⊛ **Building a Food Web**
Communities
- ❑ *community*
- ❑ *habitat*
- ❑ *niche*

Plants: The Essential Biotic Component
- ❑ *vascular plants*

Leaf Activity
- ❑ *stomata*

Photosynthesis and Respiration
- ❑ *photosynthesis*
- ❑ *chlorophyll*
- ❑ *respiration*

Net Primary Productivity
- ❑ *net primary productivity*
- ❑ *biomass*

Abiotic Ecosystem Components
Light, Temperature, Water, and Climate
Life Zones
- ❑ *life zone*

Elemental Cycles
- ❑ *biogeochemical cycles*

Oxygen and Carbon Cycles
The Nitrogen Cycle
Limiting Factors
- ❑ *limiting factor*

Biotic Ecosystem Operations
Producers, Consumers, and Decomposers
- ❑ *producers*
- ❑ *consumers*
- ❑ *food chain*
- ❑ *food web*
- ❑ *herbivore*
- ❑ *carnivore*
- ❑ *omnivore*
- ❑ *detritivore*
- ❑ *decomposers*

Ecosystems, Evolution, and Succession
Biological Evolution Delivers Biodiversity
- ❑ *biodiversity*
- ❑ *evolution*

Ecological Stability and Diversity
Agricultural Ecosystems
Climate Change
Ecological Succession
- ❑ *ecological succession*

Terrestrial Succession
- ❑ *primary succession*
- ❑ *pioneer community*
- ❑ *secondary succession*

Wildfire and Fire Ecology
- ❑ *fire ecology*

Aquatic Succession
- ❑ *eutrophication*

Earth's Major Terrestrial Biomes
- ❑ *terrestrial ecosystem*

Terrestrial Ecosystem Concepts
- ❑ *biome*
- ❑ *ecotone*
- ❑ *formation classes*

Equatorial and Tropical Rain Forest
- ❑ *equatorial and tropical rain forest*

Deforestation of the Tropics
Tropical Seasonal Forest and Scrub
- ❑ *tropical seasonal forest and scrub*

Tropical Savanna
- ❑ *tropical savanna*

Midlatitude Broadleaf and Mixed Forest
- ❑ *midlatitude broadleaf and mixed forest*

Needleleaf Forest and Montane Forest
- ❑ *needleleaf forest*
- ❑ *boreal forest*
- ❑ *taiga*

Temperate Rain Forest
- ❑ *temperate rain forest*

Mediterranean Shrubland
- ❑ *Mediterranean shrubland*
- ❑ *chaparral*

Midlatitude Grasslands
- ❑ *midlatitude grasslands*

Desert s
- ❑ *desert biomes*

Warm Desert and Semidesert
Cold Desert and Semidesert
Arctic and Alpine Tundra
⊛ **Geographic Scenes: East Greenland Photo Gallery**
⊛ **High Latitude Animals Photo Gallery**
⊛ **High Latitude Connection Videos**
- ❑ *arctic tundra*

Summary and Review
News Report and Focus Study
News Report 16.1: Experimental Prairies Confirm the Importance of Biodiversity
News Report 16.2: Exotic Species Invasion
Focus Study 16.1: Biodiversity and Biosphere Reserves

The URLs related to this chapter of *Elemental Geosystems* can be found at
http://www.prenticehall.com/christopherson

Key Learning Concepts

After reading the chapter and using the Student Study Guide, the student should be able to:

• *Define* ecology, biogeography, and the ecosystem, community, habitat, and niche concepts.
• *Explain* photosynthesis and respiration and *derive* net photosynthesis and the world pattern of net primary productivity.
• *List* abiotic and biotic ecosystem components and *relate* those components to ecosystem operations and trophic relationships.
• *Define* succession and *outline* the stages of general ecological succession in both terrestrial and aquatic ecosystems. Relate biodiversity and evolution to these concepts.
• *Define* the concepts of terrestrial ecosystem, biome, ecotone, and formation classes.
• *Describe* 10 major terrestrial biomes and *locate* them on a world map.
• *Relate* human impacts, real and potential, to several of the biomes.

Annotated Chapter Review Questions

• *Define* **ecology, biogeography, and the ecosystem, community, habitat, and niche concepts.**

1. What is the relationship between the biosphere and an ecosystem? Define ecosystem and give some examples.

The interaction of the atmosphere, hydrosphere, and lithosphere produces conditions within which the biosphere exists. This sphere of life and organic activity extends from the floor of the ocean to a height of about 8 km (5 mi) in the atmosphere. The biosphere is composed of myriad ecosystems from simple to complex, each operating within general spatial boundaries. An ecosystem is a self-regulating association of living plants and animals and their nonliving physical environment. Earth itself is an ecosystem within the natural boundary of the atmosphere. Various smaller ecosystems—for example, forests, seas, mountain tops, deserts, beaches, islands, lakes, ponds—make up the larger whole.

2. What does biogeography include? Describe its relationship to ecology.

Biogeography, essentially a spatial ecology, is the study of the distribution of plants and animals and the diverse spatial patterns they create across Earth. Ecology is the study of the relationships between organisms and their environment and among the various ecosystems in the biosphere.

Humans need to be drawn into a discussion of ecosystems and the components and cycles of the biosphere. This will add relevance for the learner and enhance an interest in the subject. Years ago I gave an environmental slide talk to a service club. The fellow sitting next to me said "business has been more difficult since they passed that ecology stuff." He seemed to think that ecology was an act of Congress or some legislature. I told him that ecology was not a law; rather, it is a fact of life, and that he is ecology—a mass of it, all alive and functioning. As Gilbert White and Mostafa Tolba remind us in the quote on page 516: "The time is ripe to step up and expand current efforts to understand the great interlocking systems of air, water, and minerals nourishing Earth." And, here we are equipped with the powerful perspective and tools of physical geography to accomplish this task!

Naked without modern conveniences, a human could survive foraging in some immediate proximity to the equator, where climates are winterless and food abundant. High productivity levels in that environment enhance survival. Yet humans occupy a wide variety of habitats worldwide, surviving in places unlike the tropical rain forests. We are ecological "generalists," detached from the environment through our wits and application of our abilities. We draw from Earth resources and subsidize our food, shelter, and clothing needs, spreading forth to every corner of Earth. Humans therefore occupy "artificial habitats," and function within a multitude of "artificial niches." Perhaps it is our detachment from most any limiting factors that creates the tremendous cost human societies are exacting from Earth.

This chapter begins with a description of basic communities and the position and operation of living species—habitat and niche—and a suggestion that the symbiosis we observe in nature is analogous to human-Earth interrelationships. Remember from Chapter 1 that the biosphere occupies that point of overlap among the atmosphere, hydrosphere, and lithosphere.

3. Briefly summarize what ecosystem operations imply about the complexity of life.

An ecosystem is a complex of many variables, all functioning independently yet in concert, with complicated flows of energy and matter.

"Life devours itself: everything that eats is itself eaten; everything that can be eaten is eaten; every chemical that is made by life can be broken down by life; all the sunlight that can be used is used. . . . The web of life has so many threads that a few can be broken without making it all unravel, and if this were not so, life could not have survived the normal accidents of weather and time, but still the snapping of each thread makes the whole web shudder, and weakens it. . . . You can never do just one thing: the effects of what you do in the world will always spread out like ripples in a pond." (Friends of the Earth and Amory Lovins, The United Nations Stockholm Conference: *Only One Earth.* London: Earth Island Limited, 1972, p. 20.)

An ecosystem is composed of both biotic and abiotic components. Nearly all depend upon an input of solar energy; the few limited ecosystems that exist in dark caves or on the ocean floor depend upon chemical reactions. Ecosystems are divided into subsystems, with the biotic portion composed of producers, consumers, and decomposers. Gaseous and sedimentary nutrient cycles comprise the abiotic flows in an ecosystem.

4. Define a community within an ecosystem.

A convenient biotic subdivision within an ecosystem is a community, which is formed by interacting populations of living animals and plants in an area. An ecosystem is the interaction of a community with the abiotic physical components of its environment. Many communities are included in an ecosystem. For example, in a forest ecosystem, a specific community may exist on the forest floor, whereas another functions in the canopy of leaves high above. Similarly, within a lake ecosystem, the plants and animals that flourish in the bottom sediments form one community, whereas those near the surface of the lake form another. A community is identified in several ways: by the physical appearance of the community, the number of species and the abundance of each, and the trophic (feeding) structure of the community.

5. What do the concepts of habitat and niche involve? Relate them to some specific plant and animal communities.

Within a community, two concepts are important: habitat and niche. Habitat is the specific physical location of an organism, the place in which it resides or is biologically suited to live. In terms of physical and natural factors, most species have specific habitat parameters (with definite limits) and a specific regimen of sustaining nutrients. Niche refers to the function, or occupation, of a life-form within a given community; it is the way an organism obtains and sustains its living. An individual species must satisfy several aspects in its niche; among these are a habitat niche, a trophic (food) niche, and a reproductive niche.

6. Describe symbiotic and parasitic relationships in nature. Draw an analogy between these relationships and human societies on our planet. Explain.

Some species have symbiotic relationships, or arrangements that mutually benefit and sustain each organism. For example, lichen (pronounced "liken") is made up of algae and fungus. The algae are the producers and food source, and the fungus provides structure and support. Their mutually beneficial relationship allows the two to occupy a niche in which neither could survive alone. Lichen developed from an earlier parasitic relationship in which the fungi broke into algae cells directly. Today the two organisms have evolved into a supportive harmony. Some scientists are questioning whether or not our human society and Earth constitute a global form of a symbiotic relationship.

• *Explain* photosynthesis and respiration and *derive* net photosynthesis and the world pattern of net primary productivity.

7. Define a vascular plant. How many species are there on Earth?

Vascular plants have internal fluid and material flowing through their tissues through specialized conducting systems. At present there are almost 250,000 species of them. This great diversity and complexity compounds the difficulty of classification for convenient study.

Land plants and land animals became common about 430 million years ago (end of the Silurian Period), with plants as sophisticated as a tree dating to about 400 million years ago (beginning of the Devonian Period). Swampy forests formed in the Carboniferous Period (280 to 345 m.y.a.) and, as we learned in Chapter 8, the continents were migrating toward the collision that formed Pangaea. The equatorial location of this continental mass is important, for it placed these lands within the influence of warm and wet equatorial climates (*rain forest Af* climates).

Conifers (gymnosperms, or cone bearing trees in which seeds are not enclosed in an ovary structure) date from the Permian Period (280 m.y.a. to 225 m.y.a.), and were followed by the flowering plants (angiosperms, or those plants with seeds enclosed in an ovary or fruit) which date from the middle Cretaceous Period (110 m.y.a.). Angiosperms quickly succeeded and are the dominant type of plant on Earth today. The animal kingdom at this time was dominated by a diverse assemblage of reptiles from very small to the largest animals which ever lived on Earth, filling every conceivable niche and habitat.

Approximately 2 billion years ago, simple splitting of cells in which genetic information is equally distributed in two daughter cells began, a process known as *mitosis*. The rise of oxygen levels in the atmosphere coincides with the appearance of more complex chromosomes and nuclear material within the cell. Approximately one billion years ago, *meiosis*, or the more complicated sexual reproduction, involving special reproductive cells, appeared in primitive plants and animals. Male and female identities to plant pollen grains (male) and eggs in an ovarian base (female) in flowers mark the dramatic arrival of sex in plants.

In thinking this through, imagine that at some time in the past every carbon atom in your body, or oxygen atom in that breath you just inhaled, cycled through a plant somewhere on the face of Earth! Life is remarkable in its ability to transmit identity and function across the span of Earth's history; some algae cells today are quite similar to the fossil remains from the beginning of the living atmosphere.

8. How do plants function to link the Sun's energy to living organisms? What is formed within the light-responsive cells of plants?

The largest concentration of light-sensitive cells rests below the upper layers of the leaf. These are called chloroplast bodies, and within each resides a green, light-sensitive pigment called chlorophyll. Within this pigment, light stimulates photochemistry. Photosynthesis unites carbon dioxide and oxygen (derived from water in the plant) under the influence of certain wavelengths of visible light, subsequently releasing oxygen and producing energy-rich organic material.

9. Compare photosynthesis and respiration and the derivation of net photosynthesis. What is the importance of knowing the net primary productivity of an ecosystem and how much biomass an ecosystem has accumulated?

The difference between photosynthetic production and respiration loss is called net photosynthesis. The amount varies, depending on controlling environmental factors such as light, water, temperature, soil fertility, landforms, and the plant's site, elevation, and competition from other plants and animals. The net photosynthesis for an entire community is its net primary productivity. This is the amount of useful chemical energy (biomass) that the community "fixes" in the ecosystem. Biomass is the net dry weight of organic material; it is biomass that feeds the food chain. Net productivity is generally regarded as the most important aspect of any type of community. Greater biomass supports greater biodiversity (more trophic levels), which results in greater stability in an ecosystem.

10. Briefly describe the global pattern of net primary productivity.

The net photosynthesis for an entire community is its net primary productivity. Primary productivity is mapped in terms of fixed carbon per square meter per year. See Figure 16.5 for a portrayal of net primary productivity. Table 16.1 lists various ecosystems, their net primary productivity per year, and an estimate of net total biomass (primary production) worldwide. Net primary productivity is estimated at 170 billion tons of dry organic matter per year.

• *List* abiotic and biotic ecosystem components and *relate* those components to ecosystem operations and trophic relationships.

11. What are the principal abiotic components in terrestrial ecosystems?

The pattern of solar energy receipt is crucial in both terrestrial and aquatic ecosystems. Solar energy enters an ecosystem by way of photosynthesis, with heat dissipated from the system at many points. The duration of Sun exposure is the photoperiod. Air and soil temperatures determine the rates at which chemical reactions proceed. Operations of the hydrologic cycle and water availability depend on rates of precipitation/evaporation and their seasonal distribution. Water quality is important—its mineral content, salinity, and levels of pollution and toxicity.

Sunflowers rotate their heads and actually track the Sun across the sky. Poinsettias flower when days are short, whereas wheat flowers when days are long and nights are short. It may seem surprising, but crops grown in arctic latitudes respond well to the long summer days. Cabbages and other vegetables grow to larger sizes than those grown at lower latitudes. Certain crops actually mature more rapidly during the longer days: beans grow faster in Alaska than in Texas. Of course, growing seasons are much shorter and the threat of frost is more prevalent.

Photosynthetic rates are only slightly affected by temperature; a range of temperatures between $10°C$ to $35°C$ is optimum for plant productivity. However, photosynthesis is sharply reduced at high air and soil temperatures, and temperatures do enhance seed germination and fruit ripening. The duration of temperature exposure seems to be more important than minimum or maximum extremes. As you move poleward, the diurnal (daily) and seasonal variability grows in magnitude and becomes an increasingly important factor. Also, mountainous topography, both in slope and elevation, is inhibiting to plant growth due to colder temperatures and exposure variability. Plants in colder temperature regions have lower photosynthetic rates and are able to function at temperatures which approach freezing. Remember though, that below 4°C (39°F), water begins to expand as it cools and upon freezing can break cell walls and damage plants. As temperatures increase, photosynthetic reaction rates increase accordingly, although leaf temperatures over 43°C (109°F) are deadly to most plants. Temperature is an important influence on physiological plant processes, including respiration. Increases in respiration reduce net productivity. Respiration releases carbon dioxide to the environment, causing one of the impacts of an enhanced global warming due to the "greenhouse effect," an increase in the atmosphere's carbon dioxide load from plant respiration.

The discussion in Chapter 7 defining the principal components of a water budget made clear the important role of water in the environment. Potential evapotranspiration represents an accurate analog of potential plant growth if adequate water is available. Farmers must manage water carefully both in quantity and timing to maximize net productivity in their crops. Plants use water to carry sugars, dissolved nutrients, and wastes throughout plant tissues. Typically, plants are about 70 percent water by weight.

12. Describe what Alexander von Humboldt found that led him to propose the life-zone concept. What are life zones? Explain the interaction among altitude, latitude, and the types of communities that develop.

Alexander von Humboldt (1769–1859) described a distinct relationship between altitude and plant communities, his life zone concept. As he climbed in the Andean mountains, he noticed that the experience was similar to that of traveling away from the equator toward higher latitudes. Each life zone possesses its own temperature, precipitation, and insolation relationships and therefore its own biotic communities. See Figure 16.7 for specific vegetation examples.

13. What are biogeochemical cycles? Describe several of the essential cycles.

Several important cycles of chemical elements operate. Oxygen, carbon, and nitrogen form gaseous cycles involving atmospheric phases. Other elements form sedimentary cycles principally involved in mineral and solid phases (such as phosphorus, calcium, and sulfur). Some elements combine the two cycles. The various processes are called biogeochemical cycles, indicating the occurrence of chemical reactions in both living (biotic) and nonliving (abiotic) spheres. The

chemical elements themselves recycle over and over again in life processes. The carbon and oxygen cycles are illustrated together because they are so closely intertwined through, for example, photosynthesis and respiration (Figure 16.8).

The key link in the nitrogen cycle is nitrogen-fixing bacteria, which live principally in the soil and are associated with the roots of certain plants—for example, legumes such as clover, alfalfa, soybeans, peas, beans, and peanuts. The bacteria reside in nodule colonies on the legume roots and fix the nitrogen from the air in the form of nitrates (NO_3) and ammonia (NH_3). The nitrogen in the organic wastes of these organisms is from a different type of bacterium that denitrifies wastes, recycling nitrogen back to the atmosphere.

14. What is a limiting factor? How does it function to control the spatial distribution of plant and animal species?

The term limiting factor identifies the one physical or chemical component that most inhibits biotic operations, through its lack or excess. A few examples include: the low temperatures at high elevations, the lack of water in a desert, the excess water in a bog, the amount of iron in ocean surface environments, the phosphorus content of soils in the eastern United States or at elevations above 6100 m (20,000 ft), where there is a general lack of active chlorophyll. In most ecosystems, precipitation is the limiting factor, although variation in temperatures and soil characteristics certainly affect vegetation patterns. Each organism possesses a range of tolerance for each variable in its environment.

15. What roles do producers and consumers play in an ecosystem? What are the detritivores? Describe their "job" in an ecosystem.

Producers and consumers are part of the food chain within ecosystems. Energy flows from producers, which manufacture their own food, to consumers, who feed on producers or consumers who are at lower trophic levels. Because producers are always plants, the primary consumer is called an herbivore, or plant eater. A carnivore is a secondary consumer and primarily eats meat (primary consumers). A consumer who eats both producers (plants) and consumers (meat) is called an omnivore. A tertiary consumer eats primary and secondary consumers, and is referred to as the "top carnivore" in the food chain. Detritivores (detritus feeders and decomposers) renew the entire system by releasing simple inorganic compounds and nutrients by breaking down organic materials.

16. Describe the relationship among producers, consumers, and decomposers in an ecosystem. What

is the trophic nature of an ecosystem? What is the place of humans in a trophic system?

From producers, that manufacture their own food, energy flows through the system along a circuit called the food chain, reaching consumers and eventually decomposers. Ecosystems are generally structured as a food web, a complex network of interconnected food chains, in which consumers participate in several different food chains. Primary consumers feed on producers and always are plants, so the primary consumer is called an herbivore, or plant eater. A carnivore is a secondary consumer and primarily eats meat for sustenance. A tertiary consumer eats secondary consumers and is referred to as the *top carnivore* in the food chain. A consumer that feeds on both producers (plants) and consumers (meat) is called an omnivore—a role occupied by humans, among others. The decomposers are the final link in the chain; they are the microorganisms—bacteria, fungi, insects, worms, and others—that digest and recycle the organic debris and waste in the environment.

An examination of the summertime distribution of various populations in specific grassland and temperate forest ecosystems reveals a stepped population pyramid. You find a decreasing number of organisms at each successive trophic level. The base of the temperate forest pyramid is narrow, however, because most of the producers are large, highly productive trees and shrubs, which are outnumbered by the consumers they can support in the chain.

Ever since Rachel Carson's book, *The Silent Spring*, appeared in 1962, there has been growing concern about the biological amplification of chemicals from pesticides, herbicides, and toxic wastes in the food chain. And even though DDT has been banned for almost 20 years, it has been legal for export since March 1981 following a presidential lifting of the ban on exporting banned products. Of course, such long-lasting chemicals end up back in the United States with imported vegetables and other food products. We really are all in this together.

A fascinating and enlightening exercise is to go back and read the newspaper and news magazine articles of 1960–1962 to see the criticisms by corporate interests heaped upon Rachel Carson's research and writing. A CBS television documentary, *The Silent Spring of Rachel Carson*, is still available. You find that the same scripts the chemical and petroleum industries used against her are being used today—playing on the theme of uncertainty and lower-threshold risk assessment myths. Imagine how an improved public learning curve would impact these repeated old arguments against acting for public trust and safety!

Not only that, but a by-product of the application of a chemical in the first place simplifies the ecosystem and reduces population diversity, which in itself introduces a level of instability in the particular ecosystem. See: William M. Stigliani, *et al.*, "Chemical Time Bombs—Predicting the Unpredictable," *Environment*, 33, no. 4, May 1991: 4–9+. Note the graph on page 7 showing Herman Daly's calculation of U.S. per capita gross national product (today assessed as the gross domestic product) adjusted for the costs of environmental damages!

An additional disturbing fact is the inadequate testing given many of the chemicals in use. A government survey in 1985 showed that 62 percent of agricultural chemicals had never been tested for cancer-causing potential in humans, 60–70 percent were not tested for possible implication in birth defects, and 93 percent not been tested as a possible cause of genetic mutations in humans. The case can be made that a vast experiment is underway in the environment—one that is testing the basic resilience of communities and ecosystems. I am baffled constantly by the fact that with possible genetic and health effects on babies and the unborn that we do not see more unity among all political factions on these issues of environmental protection!

———————————

• *Define* succession and *outline* the stages of general ecological succession in both terrestrial and aquatic ecosystems. Relate biodiversity and evolution to these themes.

17. Referring back to Figure 1.1.1 in Focus Study 1, define the scientific method. Construct a simple model to show the progressive stages that led to the development of a theory, such as the theory of evolution.

The scientific method is an application of common sense in an organized and objective manner. A scientist observes, makes a general statement to summarize the observations, formulates a hypothesis, conducts experiments to test the hypothesis, and develops a theory and governing scientific laws.

Rowland, Molina, and Crutzen wanted to know what was depleting the ozone (O_3) layer. They observed a decline in the amount of ozone in the stratosphere and hypothesized that CFCs and other chlorine containing products were depleting ozone. They collected data through surface, atmosphere, and satellite measurements and confirmed that CFCs were indeed destroying ozone. The Royal Swedish Academy of Sciences awarded them the Nobel Prize for Chemistry in 1995 for their work.

18. In what ways did the process of evlolution and natural selection contribute to Earth's biodiversity?

A critical aspect of ecosystem stability is biodiversity, or species richness of life (a combination of *bio*logical and *diversity*). The more diverse the species population (both in number of different species and quantity of each species), the species genetic diversity (number of genetic characteristics), and ecosystem and habitat diversity, the better the risk is spread over the entire community. The greater the biodiversity within an ecosystem, the more stable and resilient it is, and the more productive it will be.

Evolution states that single-cell organisms adapted, modified, and passed along inherited changes to multicellular organisms. The genetic makeup of successive generations was shaped by environmental factors, physiological functions, and behaviors that created a greater rate of survival and reproduction that were passed along through natural selection.

New mixes in the *gene pool*, or all genes possessed by individuals in a given population, involve mutations. Geography comes into play, as spatial effects are important. A species may disperse through migration to different environments, or may be separated by natural *vicariance* (fragmentation of the environment) events that establish natural barriers. Such isolation allows natural selection to evolve new sets of genes. Imagine the migrations across ice bridges or land connections at times of low sea level, or the slowly evolving rise of a mountain range, or the formation of an isthmus such as Panama connecting continents and separating oceans. To restate, the physical and chemical evolution of Earth's systems is closely linked to the biological evolution of life.

19. What is meant by ecosystem stability?

In a given ecosystem, a community moves toward maximum biomass and relative stability. However, inertial stability, the tendency for birth and death rates to balance and the composition of species to remain stable, does not necessarily foster resilience, or the ability to recover from change.

20. How does ecological succession proceed? What are the relationships that exist between existing communities and new, invading communities?

Each ecosystem is constantly adjusting to changing conditions and disturbances in the struggle to survive. The concept of change is key to understanding ecosystem stability. Ecological succession occurs when different communities of plants and animals (usually more complex) replace older communities (usually simpler). Each temporary community of species modifies the physical environment in a manner suitable for the establishment of a later set of species. Changes were thought to move toward a more stable and mature condition, which is optimum for a specific environment. This theoretical end product in an area is

called the ecological climax, with plants and animals forming a climax community—a stable, self-sustaining, and symbiotically functioning community with balanced birth, growth, and death.

Modern research has determined that in the intermediate stages of succession the greatest mineral and biomass inventories are in place. Ecosystems do not progress to some static equilibrium conclusion; such is the great diversity of nature. This is the approach of dynamic ecology. At such times of non-equilibrium transition, the interrelationships among species produce elements of chance, and species having an adaptive edge will succeed in the competitive struggle for light, water, nutrients, space, time, reproduction, and survival. Thus, the succession of plant and animal communities is an intricate process with many interactive variables, both in space and time and from internal and external processes.

Succession often requires an initiating disturbance, such as strong winds or a storm, or a practice such as prolonged overgrazing. Early communities may actually inhibit the growth of other species, but when existing organisms are disturbed or removed, new communities can emerge. At such times of transition, the interrelationships among species produce elements of chance, and species having an adaptive edge will succeed in the competitive struggle for light, water, nutrients, space, time, reproduction, and survival. When two or more species compete for limited food and resources, the more efficient organism will succeed and the other will fail in the principle of competition, or competitive exclusion. The principle of individuality holds that each species succeeds in its own way, according to its own requirements, relating to both the physical environment and the complex interactions with other species. Thus, the succession of plant and animal communities is an involved process with many interactive variables.

The stability of many ecosystems is under stress due to increasing temperatures. The key question for scientists is now how fast plants can adapt to new conditions in given habitats, or migrate to remain within their specific habitat conditions. A serious question is being asked about the possibility of plant migrations (Figure 16.17). A volume addresses many aspects of this question of climate change and plant adaptations and migrations—an induced succession: Peters, R. L. and T. E. Lovejoy. eds. *Global Warming and Biological Diversity.* New Haven: Yale University Press, 1992. (ISBN: 0-300-05056-9; 1-203-432-0940). Many different species and succession forecasts are presented with maps and discussion. See: Stuart L. Pimm and Andrew M. Sugden, "Tropical Diversity and Global Change," *Science,* 263, No. 5149, February 18, 1994: 933–34. Three maps are presented showing tree turnover rates, rates that may have profound consequences on the number of species in the ecosystem.

See: Edward O. Wilson, *The Diversity of Life*, Cambridge, MA: The Belknap Press of Harvard University, 1992, for insight into the great extinctions, an evolving biodiversity through times past, and the human impact that threatens biodiversity. His book includes a useful glossary and extensive references.

21. Discuss the concept of fire ecology in the context of the Yellowstone National Park fires of 1988. What were the findings of the government task force?

Over the past 50 years, fire ecology has been the subject of much scientific research and experimentation. Today, fire is a natural component of most ecosystems and not the enemy of nature it once was popularly considered to be. In fact, in many forests, undergrowth is purposely burned in controlled "cool fires" to remove fuel that could enable a catastrophic and destructive "hot fire." Ironically, when fire suppression and prevention strategies are rigidly followed, they can lead to abundant undergrowth accumulation, which allows the potential total destruction of a forest by a major fire. Fire ecology imitates nature by recognizing fire as a dynamic ingredient in community succession. In its final report on the fire in Yellowstone National Park, a government interagency task force concluded that: "an attempt to exclude fire from these lands leads to major unnatural changes in vegetation . . . as well as creating fuel accumulation that can lead to uncontrollable, sometimes very damaging wildfire." Thus, participating federal land managers and others reaffirmed their stand that fire ecology is a fundamentally sound concept.

22. Summarize the process of succession in a body of water. What is meant by cultural eutrophication?

A lake experiences successional stages as it fills with nutrients and sediment and as aquatic plants take root and grow, capturing more sediment and adding organic debris to the system (Figure 16.22). This gradual enrichment through various stages in water bodies is known as eutrophication. The progressive stages in lake succession are named oligotrophic (low nutrients), mesotrophic (medium nutrients), and eutrophic (high nutrients). Each stage is marked by an increase in primary productivity and resultant decreases in water transparency so that photosynthesis becomes concentrated near the surface. Energy flow shifts from production to respiration in the eutrophic stage, with oxygen demand exceeding oxygen availability. As society dumps sewage and pollution in waterways, the nutrient load is enhanced beyond the cleansing ability of natural biological processes, thus producing cultural eutrophication.

• **Define the concepts of terrestrial ecosystem, biome, ecotone, and formation classes.**

23. Describe a transition zone between two ecosystems. How wide is an ecotone? Explain.

The transition zone between two ecosystems is called an ecotone. Boundaries between natural systems are "zones of shared traits," therefore they are zones of mixed identity and composition, rather than rigidly defined boundaries. A tropical savanna is a good example of an ecotone. Situated between tropical forests and tropical steppes or deserts, tropical savanna is a mixture of trees and grasses. The savanna biome includes treeless tracts of grasslands, and in very dry savannas, grasses grow discontinuously in clumps, with bare ground between them.

24. Define biome. What is the basis of the designation?

A large, stable terrestrial ecosystem is known as a biome. Specific plant and animal communities and their interrelationship with the physical environment characterize a biome. Each biome is usually named for its dominant vegetation. We further define these general biomes into more specific vegetation units called formation classes. These units refer to the structure and appearance of dominant plants in a terrestrial ecosystem, for example, equatorial rain forest, northern needleleaf forest, Mediterranean shrubland, and Arctic tundra.

Plant distributions are responsive to environmental conditions and reflect variation in climatic and other abiotic factors. Therefore, it is important to reference this discussion with Figure 6.5 and the climate classifications.

25. Distinguish between formation classes and life-form designations as a basis for spatial classification.

Interacting populations of plants and animals in an area form a community, or association of related species. Large vegetation units, the floristic component of a terrestrial ecosystem characterized by a dominant plant community, are called plant formation classes. Each formation includes numerous plant communities, and each community includes innumerable plant habitats. Within those habitats, Earth's diversity is expressed in approximately 250,000 plant species.

More specific systems are used for the structural classification of plants. Such *life-form* designations are based on the outward physical properties of individual plants or the general form and structure of a vegetation cover. These physical life-forms, portrayed in the chapter, include *trees* (larger woody main trunk, perennial, usually exceeding 3 m or

10 ft); *lianas* (woody climbers and vines); *shrubs* (smaller woody plants; branching stems at ground); *herbs* (small plants without woody stems above ground); *bryophytes* (mosses, liverworts); *epiphytes* (plants growing above the ground on other plants, using them for support); and *thallophytes*, which lack true leaves, stems, or roots (bacteria, fungi, algae, lichens).

One method of assessment of terrestrial ecosystems is a consideration of how plants differ in moisture availability requirements. These variations result in unique characteristics associated with each water-balance regime. These characteristics are important for further understanding of plant communities, formations, and distributions. Plants are classified as follows:

Plant type, moisture, and examples
• Hydrophytes (excessive moisture) seaweed, mangroves, bulrushes
• Mesophytes (alternating deficit and surplus) common field and forest plants.
• Tropophytes (seasonal vegetation) maple, birch, larch, ash, etc.
• Xerophytes (water deficit and drought) cacti, sagebrush, junipers.

_____

• *Describe* **ten major terrestrial biomes and locate them on a world map.**

26. Using the integrative chart in Table 16.1 and the world map in Figure 16.24, select any two biomes and study the correlations of vegetation characteristics, soil, moisture, and climate with their spatial distribution. Then contrast the two using each characteristic.

A specific assignment to correlate the integrative table and the world biome map.

The author raises the question in the text as to what actually remains of the natural biomes. A map of human-induced artificial landscapes remains to be completed, for we do not have a satisfactory national land inventory in the United States, and there is political and corporate resistance to such an inventory on a global scale. We can only wonder if such fears are real when we see the U.S. *Landsat* budget threatened and a serious suggestion made by a previous (1989) administration to shut it down and what the 1990s would have been like without the *Landsat* program. If geographic analysis on a large scale is to be completed, improvement in remote-sensing capabilities will be of critical importance. Construction and launch of the various components of an Earth Observation System, or EOS, introduces a new era in the monitoring of Earth's systems—the first multi-sensor satellite called *Terra* is in orbit and operating!

Table 16.2 is meant to synthesize many elements throughout this text, and is organized along the basis of Earth's ten major biomes. The biome map (Figure 16.24) and Table 16.2 should work in concert as the student proceeds through the remainder of the chapter. Figure 16.24 is included in your overhead transparency packet.

Greater description of ecotones is included in this edition. A good example of ecotone research can be found in "Landscape Analysis of the Forest-Tundra Ecotone in Rocky Mountain National Park, Colorado," Wm. L. Baker and Peter J. Weisberg, *The Professional Geographer*, November 1995, pp 361–374.

A true-color view satellite image of North and South American from 35,000 km (22,000 mi) in space is on the first page of the text. Compare the portion of landmasses shown to the map in Figure 16.22 with this remarkable composite image and see what correlations you can make.

For detailed coverage of North America, you may want to obtain a copy of *North American Terrestrial Vegetation* edited by Barbour, Michael G., and William D. Billings, Cambridge: Cambridge University Press, 1988. From their detailed map of vegetation formations (facing the title page), you will be able to see the degree of generalization used in preparing Figure 16.24.

27. Describe the equatorial and tropical rain forests. Why is the rain forest floor somewhat clear of plant growth? Why are logging activities for specific species so difficult there?

Biomass in a rain forest is concentrated high up in the canopy, that dense mass of overhead leaves with a vertical distribution of life that is dependent on a competitive struggle for sunlight. The canopy is composed of a rich variety of plants and animals. Lianas (vines) branch from tree to tree, binding them together with cords that can reach 20 cm (8 in.) in diameter. Epiphytes flourish there too: such plants as small orchids, bromeliads, and ferns that live completely above ground, supported physically but not nutritionally by the structures of other plants. The floor of the rain forest and the floor of the ocean are roughly parallel in that both are dark or dimly lit, relatively barren, and a place of fewer life-forms—although the rain-forest floor is much livelier than the seafloor. Logging is difficult because individual species are widely scattered; a species may occur only once or twice per square kilometer.

28. What issues surround the deforestation of the rain forest? What is the impact of these losses on the rest of the biosphere? What new threat to the rain forest has emerged?

More than half of Earth's original rain forest is gone. Burning is more common than logging in deforestation because of the scattered distribution of specific types of trees mentioned earlier. Fires are used to clear land for agriculture, which is intended to feed the domestic population as well as to produce cash exports of beef, rubber, coffee, and other commodities. Every year, approximately 6.1 million hectares (15.25 million acres) are thus destroyed, and more than 4 million hectares (10 million acres) are selectively logged. This total is a loss of 0.6% of tropical rain forest each year, at which rate they will be removed completely by 2050 if the destruction continues unabated.

The text presents updated information and illustrates losses on a map in Figure 16.27. The United Nations Food and Agricultural Organization (FAO) estimates that every year approximately 16.9 million hectares (41.7 million acres) are destroyed, and more than 5 million hectares (12.3 million acres) are selectively logged. This total—averaged for the period 1980-1991 in 76 countries that contain 97% of all rain forest–represents a 0.9% loss of equatorial and tropical rain forest worldwide each year (up from the previous average of 0.6%). If this destruction continues unabated, these forests will be completely removed by about A.D. 2050! By continent, forest losses are estimated at more than 50% in Africa, over 40% in Asia, and 40% in Central and South America. Brazil, Colombia, Mexico, and Indonesia lead the list of lesser-developed countries that are removing their forests at record rates, although the countries in question dispute these statistics.

Another threat to the rain-forest biome and indigenous peoples emerged in the 1990s: exploration for and development of oil reserves. U. S. oil corporations are going ahead in Yasuni National Park, Ecuador (near the equator at 77° W) with road building and drilling. One estimate of the ultimate petroleum reserve there is 1.5 billion barrels, or enough to satisfy about three months of the U.S. demand. Similar projects are being considered in Peru.

The present human assault on Earth's rain forests has put this diverse fauna and the varied flora at risk. It also jeopardizes an important recycling system for atmospheric carbon dioxide, sources of new forms of food, and potential sources of valuable pharmaceuticals and medicines.

29. What do caatinga, chaco, brigalow, and dornveld refer to? Explain.

Local names are applied to the tropical seasonal forest and scrub on the margins of the rain forest: the caatinga of northeast Brazil, chaco area of Paraguay and northern Argentina, the brigalow scrub of Australia, and the dornveld of southern Africa.

30. Describe the role of fire or fire ecology in the tropical savanna biome and the midlatitude broadleaf and mixed forest biome.

Savannas covered more than 40% of Earth's land surface before human intervention but were especially modified by human-caused fire. Fires occur annually throughout the biome. The timing of these fires is important. Early in the dry season they are beneficial and increase tree cover; if late in the season, they are very hot and kill trees and seeds. Savanna trees are adapted to resist the "cooler" fires.

Elephant grasses averaging 5 m (16 ft) high and forests once penetrated much farther into the dry regions, for they are known to survive there when protected.

The midlatitude broadleaf and mixed forest biome includes several distinct communities in North America, Europe, and Asia. Relatively lush evergreen broadleaf forests occur along the Gulf of Mexico. Northward are mixed deciduous and evergreen needleleaf stands associated with sandy soils and burning. When areas are given fire protection, broadleaf trees quickly take over.

31. Why does the northern needleleaf forest biome not exist in the Southern Hemisphere? Where is this biome located in the Northern Hemisphere, and what is its relationship to climate type?

Stretching from the East coast of Canada and the Maritimes westward to Alaska and continuing from Siberia across the entire extent of the Russia to the European Plain is the northern needleleaf forest, also called the taiga (a Russian word) or boreal forest. The Southern Hemisphere, lacking microthermal climates except in mountainous locales, has no biome designated as such. However, montane forests of needleleaf trees exist worldwide at high elevation.

32. In which biome do we find Earth's tallest trees? Which biome is dominated by small, stunted plants, lichens, and mosses?

The temperate rain-forest biome is recognized by its lush forests at middle and high latitudes, occurring only along narrow margins of the Pacific Northwest in North America, with some similar types in southern China, small portions of southern Japan, New Zealand, and a few areas of Chile. The tallest trees in the world, the coastal redwoods (*Sequoia sempervirens*), are found in this biome. Their distribution is shown on the map in Figure 16.11a. These trees can exceed 1500 years of age and typically range in height from 60 to 90 m (200 to 300 ft), with some exceeding 100 m (330 ft). Virgin stands of other representative trees—Douglas fir, spruce, cedar, and hemlock—have been reduced to a few remaining valleys in Oregon and Washington.

Tundra vegetation is low, ground level herbaceous plants and some woody plants. Representative plant species are sedges, mosses, arctic meadow grass, snow lichen, and dwarf willow. The arctic tundra is found in the extreme northern area of North America and Russia, bordering on the Arctic Ocean and generally north of the 10°C (50°F) isotherm for the warmest month.

33. What type of vegetation predominates in the Mediterranean dry summer climates? Describe the adaptation necessary for these plants to survive.

The dominant shrub formations that occupy these regions are short, stunted, and tough in their ability to withstand hot-summer drought. The vegetation is called sclerophyllous (from *sclero* for "hard" and *phyllos* for "leaf"); it averages a meter or two in height and has deep, well-developed roots, leathery leaves, and uneven low branches. Plant ecologists think that this biome is well adapted to frequent fires, for many of its characteristically deep-rooted plants have the ability to resprout from their roots after a fire.

34. What is the significance of the 98th meridian in terms of North American grasslands? What types of inventions were necessary for humans to cope with the grasslands?

In North America, tall grass prairies once rose to heights of 2 m (6.5 ft) and extended westward to about the 98th meridian, with shortgrass prairies farther west. The 98th meridian is roughly the location of the 50 cm (20 in.) isohyet, with wetter conditions to the east and drier to the west. Examples of the problems encountered are the great distances, ground-water too deep for hand-dug wells, lack of energy for pumping, lack of fencing materials, the existence of densely matted sod causing plowing difficulties, and mobile and adaptable native populations. All these conditions represented spatial problems for analysis and later resolution, that is, railroads, oil-drilling techniques for water wells, wind energy, barbed wire, John Deere's self-scouring steel plow, and, for the unfortunate Native Americans, the Colt six-shooter.

35. Describe some of the unique adaptations found in a desert biome.

Much as a group of humans in the desert might behave with short supplies, plant communities also compete for water and site advantage. Some desert plants, called ephemerals, wait years for a rainfall event, at which time their seeds germinate quickly, develop, flower, and produce new seeds, which then rest again until the next rainfall event. The seeds of some xerophytic species open only when fractured by the tumbling, churning action of flash floods cascading down a desert arroyo, and of course such an event produces the moisture that a germinating seed needs.

Desert plants employ other strategies for survival: long, deep tap roots (example, the mesquite); succulence (that is, thick, fleshy, water-holding tissue such as that of cacti); spreading root systems to maximize water availability, waxy coatings and fine hairs on leaves to retard water loss; leafless conditions during dry periods (example, palo verde and ocotillo); reflective surfaces to reduce leaf temperatures; and tissue that tastes bad to discourage herbivores.

The creosote bush (*Larrea tridentata*) sends out a wide pattern of roots and contaminates the surrounding soil with toxins that prevent the germination of other creosote seeds, which are possible competitors for water. When a creosote bush dies, surrounding plants or germinating seeds work to occupy the abandoned site, but they must rely on infrequent rains to remove the toxins.

36. What is desertification (review from Chapter 12 and this chapter)? Explain its impact.

We are witnessing an unwanted expansion of the desert biome. This is due principally to poor agricultural practices (over-grazing and inappropriate agricultural activities), improper soil-moisture management, erosion and salinization, deforestation, and the ongoing climatic change. A process known as desertification is now a worldwide phenomenon along the margins of semiarid and arid lands. The role of global climate change and the reduction of soil moisture in marginal lands is a subject of active study with distinct connections still elusive.

37. What physical weathering processes are specifically related to the tundra biome? What types of plants and animals are found there?

Intensely cold continental polar air masses and stable high-pressure conditions dominate winter. A growing season of sorts lasts only from 60 to 80 days, and even then, frosts can occur at any time. Vegetation is fragile in this flat, treeless world; soils are poorly developed periglacial surfaces, which are continually underlain by permafrost. In the summer months only the surface horizons thaw, thus producing a mucky surface of poor drainage. Roots can penetrate only to the depth of thawed ground, usually about a meter. The surface is shaped by freeze-thaw cycles that create permafrost and frozen ground phenomena and gelifluction (solifluction) processes discussed in Chapter 14.

• **Relate human impacts, real and potential, to several of the biomes.**

38. What is the relationship between island biogeography and biosphere reserves? Describe a biosphere reserve. What are the goals?

Setting up formal natural reserves called biosphere reserves at continental sites involves principles of island biogeography. Island communities are special places for study because of their spatial isolation and the relatively small number of species present. They resemble natural experiments because the impact of individual factors, such as civilization, can be more easily assessed on islands than they can over larger continental areas. According to the equilibrium theory of island biogeography, developed by biogeographers R. H. MacArthur and E. O. Wilson, the number of species should increase with the size of the island, decrease with increasing distance from the nearest continent, and remain about the same over time, even though composition may vary. These considerations are important to establishing the optimum dimensions for biosphere reserves. The race is on between setting aside tracts of land in reserves and the permanent loss of remaining natural biomes.

The goal of biosphere reserves is to preserve species diversity. The intent is to establish a core in which genetic material is protected from outside disturbances. Ultimately, the goal is to establish at least one reserve for each of the 194 distinctive biogeographical communities presently identified. This way species diversity can be ensured, allowing us to further research and catalogue species that could prove to be medically useful, or successfully adapt to climatic change.

39. Compare the map in Figure 16.24 with the global climate map in Chapter 7, Figure 7.5. Select several regions. What correlations can you make between them?

Comparison by the student of these two maps, biome by biome could stimulate class discussions about relation to climate regions.

Overhead Transparencies

281.	16.2 a, b	Abiotic and biotic components of ecosystems (top); turtles on log in subtropical swamp (bottom)
282.	16.4	Photosynthesis and respiration determines plant growth
283.	16.5a	Worldwide net primary productivity map
284.	16.6 a	How temperature and precipitation affect ecosystems
285.	16.7	Vertical and latitudinal zonation of plant communities in life zones
286.	16.8	Carbon and oxygen cycles in the environment
287.	16.9	The nitrogen cycle
288.	16.10 a, b, c	Nitrogen map (top); satellite image of Gulf Coast waters and "bloom" (middle); map of Gulf Coast "dead zone" (bottom)
289.	16.11 a, b	Limiting factors affect the distribution of plants and animals
290.	16.13	Antarctic food web
291.	16.19 Pt. 1	Satellite image of Mount St. Helens and locator map
292.	16.19 Pt. 2 a – f	Five comparison photo pairs of successional recovery
293.	16.22a, b, c, d	Lake-bog succession (left); lake, succession in a mountain lake, and Richmond bog, Richmond. B.C.
294.	16.23	The 10 major global terrestrial biomes, world map with color key legend
295.	16.4 a	The three levels of a rain forest canopy
296.	16.6 a – f	Brazil's deforestation, 1975, 1986, 1992 (satellite images), and 2001 satellite image; ground level fire destruction; and North America map for comparison of forest lost
297.	N.R. 16.2.1a, b, c	Exotic species, examples

17

Earth , and the Human Denominator

We come to the end of our journey through the pages of *Elemental Geosystems* and an introduction to physical geography. A final chapter seems appropriate, given the dynamic trends that are occurring in Earth's physical systems.

The Human Count and the Future section has been updated. It deals with the impact of population in consumption per person and in absolute numbers. Earth's population now exceeds 6,500,000,000 people and we are increasing at over 77,000,000 per year! India and China together now total over 2.41 billion people.

The Need for International Cooperation section is as relevant as when it was introduced. It now includes the Russian ratification of the Kyoto Protocol and the Arctic Climate Impact Assessment in November 2004. The stage was set in June of 1992, in Rio de Janeiro, when a first-ever global conference, the *United Nations Conference on Environment and Development (UNCED)*, called the Earth Summit, took place. The purpose was to address with substance the many fundamental issues that relate to achieving sustainable world development. This section reviews the historic conference and a summary of the five principal accomplishments.

The agenda for these precedent-setting meetings included many topics relevant to physical and cultural/human geography. The five agreements written at the Earth Summit included the *Climate Change Framework*; the *Biological Diversity Treaty*; the *Management, Conservation, and Sustainable Development of All Types of Forests*; the *Earth Charter* (a nonbinding statement of 27 environmental and economic principles); and *Agenda 21 for Sustainable Development*.

The conference addressed protection of the atmosphere (climate change, depletion of the ozone layer, and trans-boundary air pollution); land resources (combating deforestation, soil loss, desertification, and drought); freshwater resources, and oceans, seas, and coastal areas; and the rational use and development of living resources. I think you can see that many aspects of physical geography were at the heart of this conference and subsequent conferences held during the 1990s on population, women's rights, children's rights, and energy and carbon dioxide issues.

The rest of the agenda related to other aspects of human activity, including toxics, biotechnology, species diversity, international traffic of wastes and toxic products, quality of life and the human condition, living and working conditions, and human population growth and per capita impact. Attendance in Rio exceeded 40,000!

A product of the Earth Summit was the United Nations *Framework Convention on Climate Change* (FCCC), which led to a series of *Convention of the Parties* (COP) meetings. The *Kyoto Protocol*, agreed to in 1997 to reduce global carbon emissions, was finalized in Marrakech, Morocco, in 2001 at *COP-7*. Canada ratified the climate treaty in 2002.

Subsequent COP meetings were held in New Delhi, Milan, Buenos Aires, and *COP-11* in Montreal, December 2005. Following Russian ratification, the Kyoto Protocol and Rulebook became international law in March 2005, without United States or Australian participation. Asking whether the Earth Summit "succeeded" or "failed" is the wrong question. The occurrence of this largest ever official gathering is a remarkable accomplishment. From the Earth Summit emerged a new organization—the U.N. Commission on Sustainable Development—to oversee the promises made in the five agreements. This momentum led to a second Earth Summit in 2002 (see **http://www.earthsummit2002.org/**) in Johannesburg, South Africa.

We can use these conferences as a beginning to highlight specific applied topics within physical geography.

Outline Headings

The first-, second-, and third-order headings that divide Chapter 17 serve as an outline. The students have this same outline in their *Student Study Guide.*

The Human Count and the Future
An Oily Bird
2005—A Year for the Record Books—Lessons Learned?
The Need for International Cooperation
Twelve Paradigms for the 21st Century
Who Speaks for Earth?
Review Questions and Critical Thinking
News Report
High Latitude Connection 17.1: Report from Reykjavik—Arctic Climate Impact Assessment

---

The URLs related to this chapter of *Elemental Geosystems* can be found at
http://www.prenticehall.com/christopherson

Key Learning Concepts

After reading the chapter and using this study guide, the student should be able to:

• *Determine* an answer for Carl Sagan's question, "Who speaks for Earth?"
• *Describe* the growth in human population and *speculate* on possible future trends.
• *Analyze* "An Oily Bird" and *relate* your analysis to energy consumption patterns in the United States and Canada.
• *Recall* climate change and intense weather that is occurring and *relate* them to physical geography/Earth systems science (geosystems)
• *List* the twelve paradigms for the 21st century.
• *Appraise* your place in the biosphere and *realize* your physical identity as an Earthling.

Annotated Critical Thinking and Learning

1. What part do you think technology, politics, and future thinking should play in science courses?

This is an important question, especially when considering whether or not such a capstone chapter should be included in a physical geography text. The answer for me was found in the 40-year-old quote from Marston Bates "that we are part of the system of nature." Admittedly, we have to be sensitive to the core content of physical geography, yet I think it is important that students can identify their dependent role in the biosphere and impact on natural systems. Many of our students take this class as their only science course, and this represents their only exposure to the spatial implications of the ongoing experiment Earth's cultures are conducting on this planet. This is especially important because of the specific global impact of Americans and Canadians. Gilbert White has been a leader all these years in the application of physical geography to the real world (see his quote at the beginning of Chapter 16).

2. Assess population growth issues: the count, the impact per person, and future projection. What strategies to you see as important?

Personal analysis and response. The issue in the more developed countries is not the population count but rather the impact per person. Note the simple equation in the text

$$Planetary\ Impact = P \bullet A \bullet T$$

that expresses this impact concept. Note the doubling times in Table 17.1. The issue in the less developed countries is the count and the growth of the count. As the LDC economies grow, issues of the planetary impact of those enormous populations looms as the significant concern in the new century.

3. According to the discussion in the chapter, what worldwide-scale factors led to the *Exxon Valdez* accident? Describe the complexity of that event from a global perspective. In your analysis, examine both supply-side and demand-side issues, as well as environmental and strategic factors.

On 24 March 1989, in Prince William Sound off the southern coast of Alaska, in clear weather and calm seas, a single-hulled supertanker operated by Exxon Corporation, an international energy corporation, struck a reef that was outside the normal shipping lane. The tanker spilled 41.64 million liters (11 million gallons) of oil. It took only 12 hours for the *Exxon Valdez* to spill its contents, yet a reasonable cleanup will take years and billions of dollars. Because contingency emergency plans were not in place, and promised equipment was unavailable, response by the oil industry took 10 to 12 hours to activate, about the same time that it took the ship to empty.

The immediate effect on wildlife was contamination and death, but the issues involved are bigger than these damaged ecosystems. Many factors influence our demand for oil. Well over half of our imported oil goes for transportation.

The death toll for animals was massive: at least 3000 sea otters killed (or about 20% of the resident otters), 300,000 birds, and uncounted fish, shellfish, plants, and aquatic microorganisms. Sublethal effects, namely mutations, now are appearing in fish.

This latter side effect of the spill is serious because salmon fishing is the main economy in Prince William Sound, not oil. Conflicting scientific reports emerged in 1993 as to long-term damage in the region—scientific studies commissioned by Exxon disagreed with scientific studies prepared outside the industry.

Many factors influence our demand for oil. Improvement in automobile efficiency began in 1975 due to federal regulations. During the 1980s, there was a rollback of auto efficiency standards, a reduction in gasoline prices, large reductions in funding for rapid transit development, and the continuing slow demise of America's railroad network. The demand for fossil fuels was also affected by the slowing of domestic conservation programs, elimination of research for energy alternatives, such as solar and wind power, and even the political delay of a law requiring small appliances to be more energy-efficient. Conservation plans again were politically blocked in the Department of Energy in 1990 and early 1991.

By 1999, comparatively inefficient sports utility vehicles represented half of new car sales. These SUVs are classified as light trucks and are thus exempt from auto-efficiency and some pollution standards—they burn more fuel to go fewer miles, pollute more per mile driven, and are involved in a disproportionate share of accidents. A combination of waste, low prices, and a lack of alternatives has spurred the demand for petroleum.

4. Relate the content of the various chapters in this text to the integrative Earth systems science concept. Which chapters in this text help you better understand Earth–human relations and human impacts?

Personal analysis and response by the student. The discussion introducing global wind circulation in Chapter 4 and the AVHRR images of the atmospheric effects of Mount Pinatubo's eruption might be a good place to start.

An important corollary to international environmental efforts is the linkage of academic disciplines. A positive step in that direction is the Earth systems science approach illustrated in this text. Exciting progress toward an integrated understanding of Earth's formation and the operation of its physical systems is happening right now, driven by insights drawn from our remote-sensing capabilities. Never before has society been able to monitor Earth's physical geography so thoroughly. Geographic traditions of spatial analysis and our holistic approach to planetary systems are important aspects of the emerging new science. The GIS revolution is indicative of this evolving role for geography.

5. What do you see as the lessons learned from the climate and weather events of 2005 described in this text? How might some of these lessons be applied to reducing the impact of future catastrophes?

A question for personal reflection and conclusions and perhaps a group discussion.

6. After examining the list of 12 paradigm issues for the 21st century, suggest items that need to be added to the list, omitted from the list, or expanded in coverage. Rearrange the list items as needed to match your concerns.

A question for personal reflection and conclusions and perhaps a group discussion.

7. This chapter states that we already know many of the solutions to the problems we face. Why do you think these solutions are not being implemented at a faster pace?

A question for personal reflection and conclusions and perhaps a group discussion. Obviously, the lack of information being delivered to the public by media is key. The pressure is on education to deliver. It is strange to me that with the "dumbing down" of the public through the low-expectation pop culture, that we label this the "information age."

8. Who speaks for Earth?

Carl Sagan answered his question "who speaks for Earth" with this perspective:

> **We have begun to contemplate our origins: starstuff pondering the stars; organized assemblages of ten billion billion billion atoms considering the evolution of atoms; tracing the long journey by which, here at least, consciousness arose. Our loyalties are to the species and the planet. We speak for Earth. Our obligation to survive is owed not just to ourselves but also to that Cosmos, ancient and vast, from which we spring.**

(Carl Sagan, *Cosmos,* New York: Random House, 1980, p. 345. Reprinted by permission.)

We end again with "Planet of the Year" from *Time* magazine (Figure 17.10) and Carl Sagan's eloquent question and concluding answer as to "Who speaks for Earth?" We all do! Both items remain significant philosophical moments for our times.

The challenge to all of us, especially those in the leading economies that extract most of Earth's resources, is to learn, apply, and behave in a more responsible manner than has dominated the industrial revolution to date.

The best to all of us: "May we all perceive our spatial importance within Earth's ecosystems and do our part to maintain a life-supporting Earth far into the future." And may we as geography educators realize our role in the lives of the thousands of students that pass through our classrooms.

Please feel free to communicate questions, ideas, opinions, and your thoughts about the subjects in the fifth edition of *Elemental Geosystems*. We will respond, revise, correct, and update future editions of the text and this study guide and look forward to your questions and comments. If you see either of us at one of our professional meetings, please introduce yourself and share. Thanks ahead of time for any feedback you might extend us.

The best to you, your academic career, and your future—fellow Earthling. Enjoy the challenges of new millennium!

Our contact information:

Charlie Thomsen
E-mail: thomsec@arc.losrios.edu
Home Page: http://ic.arc.losrios.edu/~thomsec
American River College
4700 College Oak Drive
Sacramento, CA 95841

Robert W. Christopherson
E-mail: bobobbe@aol.com
Home Page: http://www.prenhall.com/christopherson
P. O. Box 128
Lincoln, California 95648-0128

Overhead Transparencies

| 298. | 17.4 | World population growth intervals and future projections |
| 299. | 17.6a, b | Recent worldwide oil spills; 1989 *Exxon Valdez* oil spill map |